DATE DUE

DEMCO 38-296

The MIT Guide to Science and Engineering Communication

The MIT Guide to Science and Engineering Communication

James G. Paradis and Muriel L. Zimmerman

The MIT Press
Cambridge, Massachusetts
London, England

This book was set in Sabon on the Monotype 'Prism Plus' PostScript Imagesetter by Asco Trade Typesetting Ltd., Hong Kong and was printed and bound in the United States of America.

Library of Congress Cataloging-in-Publication Data

Paradis, James G., 1942–
  The MIT guide to science and engineering communication / James G. Paradis and Muriel L. Zimmerman.
    p. cm.
  Includes bibliographical references and index.
  ISBN 0-262-16142-7 (alk. paper)
  1. Communication in science. 2. Communication in engineering.
  3. Technical writing. I. Zimmerman, Muriel L. II. Title.
  Q223.P33   1997
  808'.0665—dc20   96-34048
                                                              CIP

Contents

Preface

Communicating is essential for any professional. In science and applied science, the writing process itself helps to form knowledge and make it available to a community of peers and to a larger world. The writer and speaker mediate between the project and its audience. Associates, clients, and other users of technical information must read a vast quantity of information to keep up in any field. Good communication makes a difference.

The materials in this book are drawn from our teaching of scientific and technical communication to two different audiences. As faculty members at the University of California, Santa Barbara, and at the Massachusetts Institute of Technology, we teach communication to engineering and science majors. As trainers in seminars in industry and government, we instruct scientists and engineers in professional settings. The materials we use in this book will, we hope, bridge the gap between the university novice and the seasoned professional.

Communication remains remarkably the same task for all writers. No one fully masters the skills required to focus a topic, organize and draft a document, and edit and package the finished form. All writers begin anew with each document, and their techniques—as even the most seasoned and experienced researcher will admit—can sustain constant improvement.

Our approach in this book is to emphasize specific processes and forms that will help individual writers create documents. We recognize, however, that writing takes place in the social context of local groups and larger organizations. Most writing in science and engineering is collaborative. Coauthored documents are cycled through editing and review and then often issued with a corporate name as author. Finished documents represent the work of many people. Collaborative writing influences nearly every phase of the process.

Another nearly universal factor in writing is the computer. Writers in science and technology use word processing, electronic files, computer graphics, and databases. They store documents in files that may be reprocessed for a variety of audiences and situations. They use on-line searches to do research. They convert traditional documents to electronic forms and replace hefty multivolume manuals with a single multimedia CD-ROM. They use the information resources of the Internet and have personal as well as organizational home pages. To date, however, there is no electronic substitute for a skillful editor like Debi Osnowitz, who has worked with us on this book.

Throughout, we make a special effort to provide realistic examples from actual documents and situations. Most of our examples have been used in college classrooms and professional seminars. Our experience is the basis of our book.

Acknowledgments

I am grateful to the many teachers, colleagues, students, and clients who have taught me, read manuscripts, furnished examples, and given me advice, especially Jim Souther, Mike White, Robert Rathbone, John Kirkman, Peter Hunt, Muriel Zimmerman, John Kirsch, Ed Barrett, Marie Redmond, Harold Hanham, Anthony French, Tom Pearsall, Charles Bazerman, Charles Sides, Jim Zappen, Les Perelman, Dave Custer, Dan Cousins, Chris Sawyer-Laucanno, Bob D'Angio, Anne Lavin, Kenneth Manning, Leon Trilling, Frank McClintock, Jay Lucker, Tom Weiss, and Mary Pensyl. I appreciate the insights and concrete suggestions given me by my University of Washington students and my MIT students over the past two decades. I am also grateful to the many engineers and scientists at The Applied Physics Laboratory (University of Washington), Brookhaven National Laboratory, the Department of Interior, Department of Energy, Exxon, and Mitre Corporation for teaching me about the roles communication plays in the work of professionals. I owe my family—Judy, Em, and Roz—for their many hours of good-natured indulgence of my "writing project."

Cambridge, Massachusetts   Jim Paradis

For some years, I divided my academic life between MIT and the University of California, Santa Barbara. Students, colleagues, and friends on both coasts have contributed to my general understanding of technical communication, as well as to specific issues addressed in these chapters: among them are John Balint, Jack Falk, Randy Goetz, Stan LaBand, Mel Manalis, Kenneth Manning, Judy Messick, Joan Mitchell, Jim Paradis,

Sam Skraly, and Ellen Strenski. I need to put Hugh Marsh—colleague, collaborator, storyboarder, and valued friend—in a sentence of his own. The graphics in chapters 7 through 16 and chapter 18 were prepared by Denise Belanger, Alex Nathanson, and Gretchen Smith. In my view of the connection between communication and technological achievement, I have been strongly influenced by my father, Nathan R. Laden (1907–1977), civil engineer, who spent much of his professional life at the demanding task of specification writing, translating design concepts into plain and enforceable language. Without the advice of my husband, Everett Zimmerman, Professor of 18th Century Literature at the University of California, Santa Barbara, this book would have been longer and finished sooner.

Santa Barbara, California   Muriel Zimmerman

The MIT Guide to Science and Engineering Communication

1

Science and Engineering as Communication

Consider this situation. A research group carries on an informal discussion with colleagues and management. Through the discussion, the group develops an initial concept for a new coal atomization process. This concept is made formal in an in-house proposal to local management and then as a detailed proposal to a government or industrial sponsor. The project is funded. A few months later, the ideas are worked out in greater detail in the researchers' notebooks and databases as narratives, computer files, and data curves.

Some of this material furnishes the transparencies for Thursday afternoon in-house seminars; some developments become part of presentations at professional conferences. Later still, the same notes, data files, and figures are recorded and circulated as progress reports to sponsors. Eventually—after still more informal discussion, progress reports, and seminars—aspects of the researchers' coal atomization process take shape as one or more formal reports, journal articles, process specifications, patent applications, or design standards. Along the way, the group will have generated a good number of administrative and technical memoranda and letters.

Research and development (R&D) are carried out in a context. Written communication is the vehicle that puts ideas and information that science and engineering generate into context. Documents are the instruments of this process. Because writing and speaking skills figure prominently in the ability to communicate important aspects of scientific work, the skilled writer is typically the most effective engineer or scientist.

(a) % total job-related time

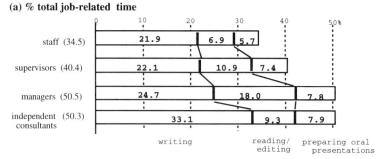

(b) % total writing-related time

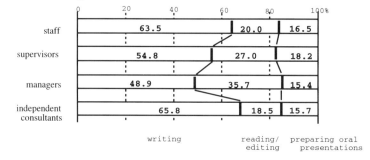

Figure 1.1
Time 265 R&D engineers and scientists estimate they spend on writing and allied activities. Note that both communication and editorial activities increase with decision-making responsibilities. (Source: Bower 1985)

Communication skills are so essential that writing often becomes a large part of any job. One survey of 265 R&D engineers and scientists working in more than 20 organizations recorded that these professionals spent at least one third of their job time on writing and related activities (Figure 1.1). Moving up the organizational ladder, to supervisor and then to manager, they spent increasingly more time editing and reviewing the writing of their subordinates as they assumed responsibility for meeting objectives and deadlines. Independent consultants spent even more time preparing documents.

In the same survey, nearly half the respondents—including 71% of the managers—said they could cite cases in which writing ability had a serious effect on a person's career. The following comments point out how writing makes a difference (Bower 1985):

"Writing ability often correlates with a superior manager's characteristics, such as the capability to synthesize and clarify in complex situations."

"The quality of communication has a strong impact on an idea's reception."

"Poor proposals and reports mean little or no future work with clients."

"Young engineers who write with clarity and make logical presentations tend to become supervisors of other engineers within 5 years of graduation."

"A person got more visible work done because she finished reports quickly, leaving more time for other work."

"An engineer was forced to resign because the research reports he wrote were poor, even though the quality of his work was very good."

"Oral skills are more important to 'power seekers,' but good writing helps the few craftsmen who are left."

Written Communication

Communication in any institution or organization is a fluid activity. Writing extends and complements other forms of human interaction. It is similar to speech, but it requires a different kind of discipline and meets different aims.

Writing freezes thought processes, research records, specifications, decisions—anything that can be represented in words, symbols, or graphics. Documents are records of the steps of decision making, design, reasoning, and research. Writing is the preeminent means of transferring information and knowledge in detail and accuracy.

Written communication does, however, take time to prepare and should not be used when a telephone conversation will do. For instance, you don't need a memo—and maybe not even e-mail—to ask the time of a meeting. The formal, permanent aspects of a document may also be inappropriate when a more personal touch is required or when a record isn't needed. The written word can be analyzed in minute detail, so your language must be clear, consistent, and accurate. These qualities come

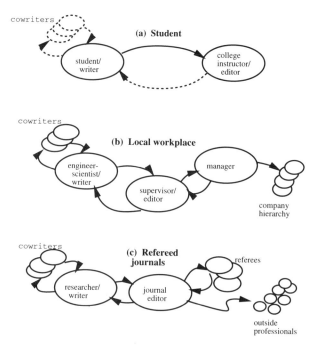

Figure 1.2
Writing is intensely cyclical, as these review cycles suggest.

not only from your facts but also from the organization, grammar, style, and mechanics of your document.

The Social Context of Scientific Writing

Scientific writing is social in two senses. First, it is typically collaborative, the result of teamwork among peers and management. Second, the written document itself circulates in a community of specialists. An internal review process helps writers shape information into useful arguments that address their projected readers. Collaborators may be colleagues, supervisors, or outside readers. They may contribute to the finished product. They may provide comments and information. Or they may guide and evaluate the work.

The reviewing process, as shown in Figure 1.2, has different implications in different environments. Student writing, for example, is rarely true collaboration and has no audience beyond the instructor. This way of learning sometimes leads the novice to underrate the importance of writing in the professional world. In-house writing, on the other hand, is generally examined by both colleagues and a supervisor, who edit for both content and style. In formal publication, the document passes outside the institution to a professional editor, who circulates it to referees and may ask for revisions.

Editorial commentary is better early than late. As writers become more involved and get further along, they frequently become more resistant to suggestions for change. Changing an outline is easier than rewriting a draft. If you wait until you have finished the document before you ask for input, the suggestions you receive may require new organizing, writing, and maybe even new research. Simple psychology suggests that you ask for advice early, when you've invested less time and emotion. People have been known to write 75 pages when only five were required.

The Politics of Written Communication

Most writing will have some political significance, quite apart from the primary message. To write is to assert, and assertions involve other people's interests. Your information may be accurate and your argument worthy, but you can still make big communication errors, as one young engineer did, in this account from a professional journal (Guterl 1984):

An engineer who was being groomed for a high position in his company sat at his desk in disbelief. Top management had rejected his proposal for installing computers in the manufacturing building. How could the managers fail to understand that computers would bring efficiency and savings? The old way had worked fine, they said, and besides, the computers would cost too much.

To dispel his managers' false notions, the engineer armed himself with facts—more facts, he was sure, than anybody else had ever known about the manufacturing department—and at a meeting, he challenged the top executives to prove him wrong. It did not work. He had embarrassed the president of the company, who knew little about computers. They not only rejected his proposal but also promoted somebody else to the position he had sought, that of director of engineering. To make matters worse, the new director promptly had the new computer system installed.

Clearly, this engineer thought himself a direct kind of person, priding himself on sticking to the facts. But the facts, whatever their technical merit, seemed confrontational to management. Rather than building consensus in individual or small-group meetings, the engineer put his manager on the spot by rushing his proposal into a document. He was pursuing personal aims, but in his apparent desire to secure credit for his idea, he denied management a role in the innovation. He then made matters worse by publically demonstrating management's ignorance.

The Importance of Digital Technologies

Writers often do not appreciate the extent to which their activities are collaborative, whether in focusing a problem, developing a document plan, or drafting, revising, and producing a manuscript. At each of these stages, a writer needs to consult colleagues and supervisors, review notes and data files, and rework the initial efforts. The computer is an essential tool for managing the work of communication.

This capacity for managing writing activities, combined with the immense capacity of computers to process information and access worldwide information sources, makes computers indispensable to writers. Computers are ideal for:

1. Breaking large tasks into smaller ones that can be worked on independently
2. Storing and processing information
3. Drafting and revising prose
4. Preparing graphics
5. Cycling the various parts of a document through the review process
6. Searching for database and library information
7. Merging the various elements to produce the final document
8. Routine communication with colleagues

Any writing project, large or small, requires several coordinated activities that transform original data into draft documents and, finally, into finished documents. As shown in Figure 1.3, a project team can accomplish all this work efficiently by storing files in computers. Computers can also help you tailor a body of material to fit the different aims and audiences of a proposal, procedure, memo, speech, graphic, progress report, or research article.

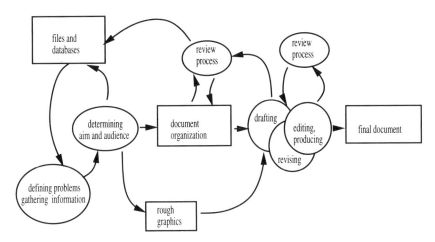

Figure 1.3
Activities associated with professional writing. Cyclical activities lend themselves well to computer-organized efforts, which also merge them effectively.

Recording as the Basis for Writing

Effective writing requires initial organization, something that writers sometimes underestimate. When information becomes available, you need to preserve it. The articles or reports you fail to file, the comments you do not record, the meeting notes you lose, the curve or data you don't get around to entering, the files you fail to organize in the computer, the procedure you forget to write down—any of these lost or neglected items can haunt the researcher-turned-writer. Even small items—a missing reference, a physical constant, a procedural description—can turn a routine writing task into a guessing game. The failure to organize information as it's gathered accounts for the great inertia many writers experience when trying to get started.

It's sometimes tempting to think that comprehensive research precedes all writing. This is clearly not the case. Numerous writing and information-gathering activities take place while research is carried out, and these activities, in turn, furnish the basis of all project-related writing.

Consider a research project in which a physicist, physician, and medical technologist conduct a 5-month series of experiments to study the pattern and extent of lithium distribution in sections of human brain. The investigators collect over 20 recent papers on lithium treatment of mania and depression, nuclear analytical procedures for analyzing lithium distribution, and modes of lithium action in rodent brain tissues. They use a high-frequency beam reactor to bombard human brain tissue samples with neutrons, which cause a lithium isotope in the brain to release energetic particles.

They fill several notebooks with the details of the experimental design, methods of preparing cross sections of brain tissue, inscription records of the cross sections, data from particle detection, data reduction and rough graphs, notes on error analysis and sensitivity ranges for the experiment, and case histories of deceased patients who had undergone lithium treatment. Funded by a national health foundation, they are expected to prepare a report and to publish two or three papers on their findings in refereed journals read by clinicians and health researchers.

Like most research projects, this one generates an immense—and potentially chaotic—volume of written and visual detail long before any formal write-up of results takes place. The detail is a combination of previously published papers, a proposal, correspondence, photographs, spreadsheets, graphs, patient records, notebooks, and notes from meetings and informal discussions. This thicket of information needs to be sorted and arranged so that its patterns can be studied and it can be retrieved when necessary.

Planning a Recording Program

A program of information gathering, recording, and archiving is a way of anticipating the written presentations that will inevitably follow. The ability to get to the various sources of information is essential to solving problems. Here are some suggestions:

• Design a computer filing system or spreadsheet that will arrange data and text files for anticipated use in writing.

• Arrange published materials, correspondence, and other collectibles in file folders, loose-leaf folders, and computer or vertical files.

• Keep a notebook of all meeting notes and agendas for future reference.

• Record experimental procedures, details, notes, and procedures in routinely updated laboratory notebooks.

- Sketch and arrange preliminary graphics in laboratory notebooks or in vertical files.

Your design for arranging and storing material will save hours later and may well save you from having to reconstruct events from an incomplete or vague record.

Using Notebooks

Although organizing records of your accumulating work may at first seem like drudgery, your records and files do assume great value with time. They are your personal store of information, extensions of your memory (Figure 1.4). Records require you to sort information conceptually. What is included and what is left out are matters of great significance.

Systematically kept, your notebook preserves the content and sequence of your activities. Your notebook makes it possible to reconstruct project developments. Always date the pages. A research record in a permanently bound notebook with printed page numbers is also a legal record of ideas, drawings, or descriptions. Items commonly recorded in notebooks include

- *Objectives*: the purpose of an experiment and the time of day of the experimental activity
- *Procedures*: rough descriptions, sketches of apparatus, modifications to apparatus, steps in the procedure, notes on equipment and materials used
- *Results*: columns of data, rough graphs, descriptions, observations, photographs, printouts
- *Analyses*: equations, narrative comments, unanswered questions, data reduction techniques, new ideas, references to the published literature (textbooks, handbooks, articles), correlations of data

You may also record experimental data and notes in digital files, but these may not be legally acceptable as original forms. Maintain vertical files for material that does not fit in the notebook. Drawings, photographs, blueprints, equipment specifications, computer printouts, and calculations are all worth saving. Project record keeping is crucial. Laboratory notebooks may be subpoenaed in court cases that concern experimental or design questions. You may be liable if you fail to maintain files of calculations, sources and grades of materials, design changes, or

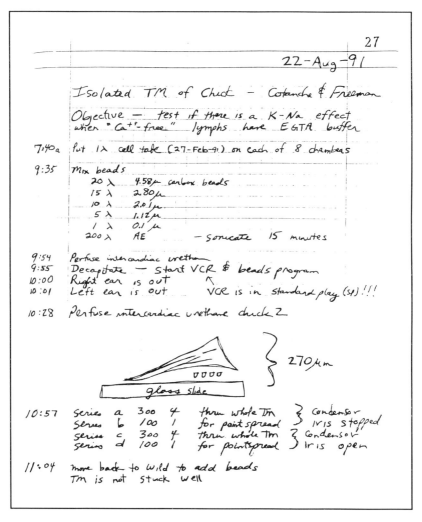

Figure 1.4
Notebook entries for experimental study of laboratory chick specimens. Note the statement of experimental objective and the linkage of time and action. (Courtesy of Professor Thomas F. Weiss, MIT)

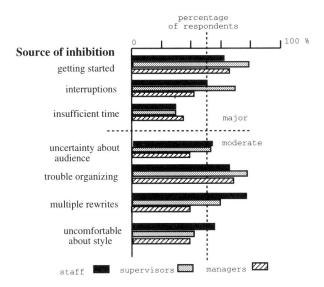

Figure 1.5
Writing inhibitors, as rated by 265 R&D engineers and scientists. Bars represent percentage of each group rating the listed factor as a major or moderate problem. (Source: Bower 1985)

any essential phase of your work. Your project files may later be inexpensively stored on floppy disks and microfiche.

Avoiding Writer's Block

The writing process is complex and abstract enough to offer many kinds of barriers. A writer who can't get started may not be able to identify the source of the problem. Any aspect of writing, if neglected, can effectively halt writers in their tracks.

In the survey of R&D scientists and engineers, inhibitors of writing were mostly time related (Figure 1.5). Respondents cited getting started, interruptions, and insufficient time as principal problems experienced by staff, supervisors, and managers. Writing inhibitors were related to both writing conditions and writing content, with writing conditions considered more serious. The lesson here is that working conditions, such as privacy and time allocation, can make an immense difference to the writer.

The problems of getting started and being interrupted indicate that many R&D writers have trouble launching their work. Their problems may be the result of inadequate recording and archiving strategy, confusion over the task required, or anxiety over time. Other inhibitors include assessing the audience, organizing the document, revising, and working on style. Organizing is a problem of staff, supervisors, and managers alike. Staff members, who are generally less experienced, also tend to have trouble with rewriting, perhaps because of having to satisfy supervisors and managers, perhaps because of poor writing skills.

Writing, as Figure 1.3 suggests, is an assortment of activities. A writer having trouble might be stuck at any of several phases. Here are some strategies to consider if you're having trouble.

• *Make sure you have enough information.* If you don't have enough information or don't have the right information, you need either to carry out more research or to change your goals.

• *Define your task specifically.* Some writers stumble because they don't have a specific aim and are unclear about their audience. If you don't address these issues at the planning stage, you will have trouble at the drafting stage.

• *Organize your material with your task in mind.* Organizing material can be difficult at first because technical information comes in different formats from many different sources. The sheer mass of source material can make sifting through and reducing the detail an essential step in your project.

• *Get feedback from colleagues and supervisors.* Circulate working outlines for discussion. Asking for feedback early can promote consensus about goals, coverage, and strategy in drafting a document. Preventing errors and misunderstandings is better than repairing documents later.

• *Prepare your rough graphics.* Developing graphics is a very effective way to focus your work. Not only are graphics critical to exposition in all fields, but they also summarize essential information. Many writers begin organizing their work by assembling graphics and then shuffling them to work out the logical sequence of their prose.

• *Organize your writing space.* Arranging your research materials and organizing your computer files can help you establish control. Anxiety over the location of materials can lead to writer's block. You're likely to need quick access to notebooks, spreadsheets, published sources, project proposals, reference works, rough drawings, note cards, and correspondence.

• *Be ready to make judgments or decisions.* At the writing stage, you're putting your views and findings on record. This act of formalizing can pose great difficulties when, as is often the case, your results are not all that clear. Remember that writing is itself a decision-making process. Don't put off writing until you've achieved some mythical level of certainty.

• *Don't try to write a perfect first draft.* A writer who expects to write a perfect first draft is just the person who spends the morning putting a comma in and the afternoon taking the comma out. If you're convinced that your writing should progress routinely through a linear series of steps, you're going to hit a wall. Assume that you'll need to rewrite.

Writing as Process

Writing requires planning, drafting, revising, editing, and producing—activities that are usually sequential. Novice writers often equate writing with drafting and proceed without much of a plan. Trying to write without a plan, however, can lead to false starts, confusing introductions, bloated documents, inappropriate material, wordiness, and incoherent organization. Think of planning as part of the writing process.

Fortunately, no one in science or engineering needs to start without a plan. In the chapters on memoranda, proposals, reports, and journal articles, you'll find formats and processes for most writing situations. As the architect of your document, you may need to modify the format, but approaching writing as a process can save a lot of time. Always think about revising, editing, and producing as parts of your writing. As you will see, these processes are essential to creating effective documents.

2

Defining Your Audience and Aims

The research scientist studying how neurons fire expects to prepare grant proposals for project support and eventually publish papers for colleagues. But neuronal firing means different things to potential funders and colleagues. The civil engineer preparing a report on soil samples at a bridge site is writing for architects, building contractors, town managers, and Environmental Protection Agency agents. Both the researcher and the engineer must consider the audience, the people who will read their writing.

Identifying the readers' needs and interests turns out to be one of the most important parts of writing. Science and engineering are problem oriented, and stating problems clearly helps focus resources on answerable questions. To keep problems from existing in purely abstract terms, a writer needs to identify the constituency interested in the problem.

The question then becomes this: What is the best strategy for meeting those readers' needs? For example, an industrial engineer might see automating a manufacturing operation as a technical problem. But it is also a financial problem that needs to be justified administratively by managerial decision makers. The writer whose proposal simply concentrates on a technical explanation fails to shape the arguments for the readers who will ultimately make the decision. This misunderstanding can defeat a writer's aim.

Determining Your Readers' Interests

Readers are usually motivated by their job responsibilities as decision makers (managers), knowledge producers (experts), operators and maintainers (technicians), and generalists (lay people). But these different audiences are abstractions or, at best, averages. Not every expert in particle physics is going to think the same way, use the same methods, or have the same problems. The veteran technician knows more about many technical subjects than the university-trained colleague.

Addressing your audience is even more complicated when the audience includes managers, specialists, technicians, and lay people. Each part of your audience will need to find the information it needs. An audience of managerial readers, for example, will analyze what you have to say in the terms of their decision making: costs, benefits, alternatives. The expert, technician, and lay reader will also analyze your message according to their interests and responsibilities.

Some documents have a primary audience, which you can often identify by clearly defining the purpose of the document. For example, if you aim to establish the laboratory procedure for preparing titanium oxide by the ethanol-water method, then you are addressing individuals with technical concerns. If, on the other hand, you aim primarily to show that titanium oxide precipitates prepared by the ethanol-water method are 32% more durable than those prepared by an alternate method, then you are speaking to experts interested in innovations. If instead you set out to argue the feasibility and economy of a 3-year $800K program to develop an industrial process for synthesizing titanium oxide by the ethanol-water method, you are writing for managers concerned with planning and resource allocation. Your aim should identify your audience.

Coverage, Organization, and Technical Level

No formula will produce writing appropriate for a given audience. Instead, you need to analyze several variables as you shape a document for a projected readership. The expectations of your audience will determine the coverage you give your subject, the organization you give your material, its technical level (including graphics), and, finally, your tone (Figure 2.1).

Your choice of material is your first decision. Coverage refers to the scope of the subject, and it can vary greatly. Let's assume that you are proposing a 3-year $800K program to develop an industrial process for

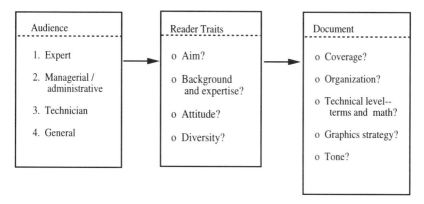

Figure 2.1
The interplay of audience, reader traits, and document criteria. Reader needs determine document criteria.

synthesizing titanium oxide by the ethanol-water method. You can cover your subject in numerous ways, depending on your readers' main concerns. These might be financing and administratively supporting the work, judging the feasibility of the process, or examining the technical elements of the project. For a managerial audience—those who must decide whether a project merits the resources it requires—you might focus on the commercial potential of titanium oxide. For technical experts, you could more fully describe the synthesizing process. For the lay reader, your focus is likely to be on the usefulness of the process and the substance to be produced.

Organize your material around your readers' priorities. For example, assume that you are a civil engineer writing to analyze the seismic site conditions of a proposed San Francisco harbor facility and to make appropriate design recommendations. You can organize the same information in two distinctly different ways, each anticipating the priorities of the specialist or the managerial reader, as shown in Figure 2.2. One version develops materials so that the technical specialist can analyze the process leading up to the actual design recommendations. This kind of organization allows critical evaluation of a knowledgeable expert. The second version emphasizes the major findings and recommendations because these are the most important information for a planner. Conclu-

Topic: Seismic Design of San Francisco Waterfront Facilities

Specialist audience	Managerial audience
Introduction (problem and background)	Summary of problem and findings
Analysis	Design recommendations
o Local geology and seismicity	o Background
o Site subsurface conditions	o Recommendations
o Potential seismic hazards	o Discussion of design
Experimental methods	recommendations
o Slope stability tests	Analysis
o Methodology	o Local geology and seismicity
o Results	o Site subsurface condition
o Earthquake-induced yard	o Potential seismic hazards
settlement tests	o Discussion of experimental results
o Methodology	Conclusions
o Results	(Appendix A) Data on slope stability
Discussion	(Appendix B) Data on earthquake-
Design recommendations	induced yard settlement

Figure 2.2
How the same material (on seismic conditions in San Francisco) could be organized for two different audiences.

sions, therefore, come first, with details given in order of likely interest. The detailed experimental information is included at the end, to support the recommendations.

Another way to tailor documents to audiences is to adjust the language, especially the special terminology, mathematics, symbols, and graphics. Assume, for example, that you want to explain the operation and use of the carbon dioxide laser scalpel. The subject deals with laser technology applied to surgery. The topic is specialized in both engineering and medicine. The audience might include hospital administrators, surgeons, patients, nurses, medical technologists, and research engineers. Considering their needs and educational backgrounds should help you determine what language is appropriate.

The following passage might be addressed to research engineers. Its fairly advanced technical level is reflected in the terminology and use of symbols, references, and units:

The CO_2 laser scalpel connects to a monotoxic optical fiber made of silver halide, originally constructed and tested at the University of Tokyo (Atsumi et al. 1983). The core of light fiber that transmits the infrared beam of the CO_2 laser is made of $AgBr_2$ and is clad with a layer of $AgCl_2$. Transmission loss in the fiber is 0.22 dB/m at 10.6 μm.

Specifying output and other operating parameters gives the reader technical information about the instrument's capabilities. References to the literature enable the user to obtain other important data. This level of technical discourse would be inappropriate for nonexpert audiences, who could not be expected to understand the language.

The same topic directed toward the surgical nurse would incorporate less technical language but would cover the operating procedures. The emphasis here is on safety and standard practice, with directly worded prose.

Although the type of laser used and the surgical applications are determined by the surgeon, the nursing staff must ensure that the equipment is regularly inspected and maintained and that potential fire hazards are avoided during operation. The assistant should drape the area adjacent to the operative field with wet towels, which should be remoistened frequently during the surgical procedure (see Figure 6). A large container of saline solution must be kept available, both to moisten the operative area and to douse flames if material ignites. Because all endotracheal tubes are subject to ignition, special precautions are necessary for a laryngoscopy with endotracheal intubation.

The technical level of the language is also influenced by the accompanying graphics, which can vary greatly for any given subject. The graphical presentation of information can range from highly specialized line graphs to general pictures.

Document Pathways in House
Another way of thinking about audience is the document pathway. You know that your report or memorandum travels, accumulating readers along the way. Most organizations have a hierarchy through which written communication passes. The organizational chart often shows some of this path. The document may well move up the hierarchy through

supervisor I, the group leader, to the division manager, and finally to the research director. Then it might travel back down to the staff level of some related group or over to marketing. Figuring out where the document will go is in a writer's interest. The document pathway will tell a lot about your audience and therefore about the general coverage and organization needed in your document.

Your understanding of this audience comes from knowing what the people reading your document will want to do. Some will act on it; others will peruse and archive it; many will just dump it in the wastebasket. Still others will use it to assess the quality of your work. The best preparation you can make for understanding your readers is to study your organization and familiarize yourself with its staff and their responsibilities.

The Peer Specialist   Assume, for example, that you are reporting on the structural and commercial advantages of a new lightweight composite for building frames for roof-installed solar panels. Several structural engineers and materials specialists might comment on your first draft. These group members share an interest in the success of the report. They read the document, add their comments in the margins, and contribute ideas. Often, one or more of them will be listed as coauthors. As an audience, then, this group is closest to the subject matter and may be the most technically informed.

The Supervisor   The group supervisor, who typically is held accountable for meeting larger corporate research objectives, has a major stake in the documents the group produces. The supervisor is highly motivated to make the document focus on the right objective in an effective way. Draft documents often fail to make a clear statement. They also fail, sometimes, to address the established aim of the group enterprise, or they don't communicate in effective and organized prose. This person may comment on the document at its first draft stage and cycle the document back for revisions (Figure 2.3). Reading and revising can be very demanding, especially for novice writers. As documents return for second or third revision, the tension can build.

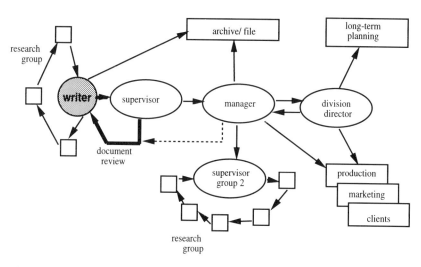

Figure 2.3
The R&D writer's rhetorical world. A document typically finds its way into several different situations in which individuals of the local organization use the information in different ways.

The Manager   R&D groups often report to a manager, who co-ordinates and directs the broad effort. This higher-level administrator does not participate, normally, in preparing individual documents but is undoubtedly one of the document's most important readers. Like group supervisors, managers release documents to meet their responsibilities for knowledge production, but they also monitor research goals and progress, a process in which the written report plays an important part.

Writing for Publication

Specialist, manager, and lay audience take on new meanings when you write for the published record. Away from the in-house R&D environment, the audience often becomes easier to identify. The large, specialized readership of a given journal, for example, can be assumed to have similar education, professional interests, and technical expertise. Most articles are written to expand the knowledge base in a given field, and an expert audience can therefore follow terminology, mathematics, and methods. You can develop communication strategies partly by inspecting the documents already produced (e.g., proposals, refereed articles, formal

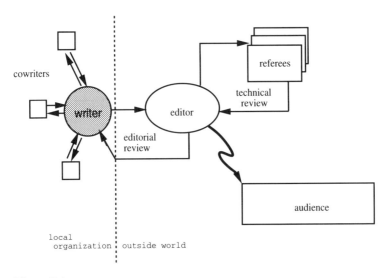

Figure 2.4
The rhetorical world for a writer contributing to a refereed journal. Although the dynamics of review are similar to those met by the R&D writer, the process is more formal and moves outside the local environment.

research reports). The coverage, language, and graphics are frequently similar throughout a journal publication.

The document pathway typically moves through an editor to several referees (Figure 2.4). The editor is not your supervisor but the employee of another organization. Note that Figure 2.4 shows two cycles. The editor acts as coordinator and stylistic arbiter for the first cycle, and referees judge the technical merits of the document in the second cycle. If the document in question is a proposal, this outside referee audience will file reports supporting or rejecting the funding. If the document is a refereed article, the referees file reports supporting or opposing publication. Often, the referees recommend revision followed by resubmission.

Framing Your Writing Project

Science and engineering are part problem solving. Defining your problem will improve your writing. The more specific your problem, the better you can design your research and determine your goals.

But problems can be approached in many different ways, depending on your audience. Developing air bags for the automobile industry poses different kinds of problems for the CEO of Chrysler, the chief of its marketing division, its air-bag design team, Ralph Nader, and the average driver. The framing of a problem depends both on identifying your audience and on determining your topic.

From Technical Problem to Writing Topic

A problem is a conflict that someone wants to remedy. For example, "the scanner on board the Landsat 4 satellite is malfunctioning" is a problem. A topic renders the problem and its solution into a focused research and writing objective. Possible topics related to the satellite malfunction are

- Causes of scanner malfunctioning on board Landsat 4
- A protocol for correcting Landsat 4 scanner malfunctioning
- An improved design for Landsat electromechanical scanners

Each topic reflects the original problem but refines and limits it. Narrowing your topic in this way defines your research and writing goals.

Your topic reflects your audience and its interests. In research environments, problems typically reflect state-of-the-art theory and practice and hence tend to be more abstract and specialized. For example:

- What is the effect of single amino acid replacement on the thermal stability of the N-terminal domain of a k-repressor?
- How can we characterize nematic ordering in lyotropic liquid crystals?
- What are the effects of iron additives on soot particle formation and growth?

Here problem definition is demanding. It takes a seasoned research engineer or scientist to identify critical questions that lie within the domain of experimental or theoretical possibility.

Engineers and scientists working in industrial R&D generally don't generate the topics they write about. They work on problems identified by clients, colleagues, research directors, or supervisors. For example:

- What is causing the recent chemical degradation of our O-ring seals?
- What design should we use in the blade of our C-53x windmill?
- Should we adopt suspension coils or air cushions to improve the ride quality of our XX-100 minibus?

• What fire-alarm system will best protect our client's warehouse installation?

Yet even these well-defined problems require refinement before you have a topic you can confidently start to work on.

Clarifying your topic provides you with some control. By defining goals within your resources, you make your success more likely. For example, you would not want your topic to include an improved design for Landsat 4 scanners if you were prepared only to describe the causes of their malfunction. Starting work with a clear topic helps you avoid committing yourself to a project with no boundaries.

To refine your topic, read background material and related project work and discuss these with colleagues, experts, clients, or supervisors. Write down what you know about the problem you're addressing. A clear problem statement, such as "What is causing the recent chemical degradation of our O-ring seals?," helps focus your investigation. Then articulate the main topic. A topic like "Lyme disease in public parks" is much less manageable than one like "Recommended measures for reducing Lyme disease at Crane's Beach." The former topic is broad and requires a wealth of information to cover; the latter is specific and restricts the scope of your research to a local effort (Figure 2.5).

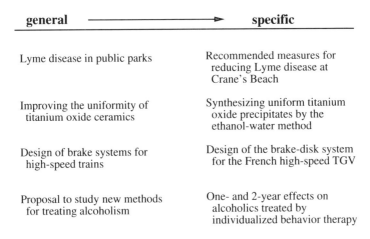

general ⟶	specific
Lyme disease in public parks	Recommended measures for reducing Lyme disease at Crane's Beach
Improving the uniformity of titanium oxide ceramics	Synthesizing uniform titanium oxide precipitates by the ethanol-water method
Design of brake systems for high-speed trains	Design of the brake-disk system for the French high-speed TGV
Proposal to study new methods for treating alcoholism	One- and 2-year effects on alcoholics treated by individualized behavior therapy

Figure 2.5
Make your topics specific. This simple move can save time at the writing stage.

Avoid becoming rigidly attached to your first problem definition. Problems are likely to evolve with information gathering. Experiments, reflection, and discussion with colleagues will help you refine your problem. Initial assumptions are often quite wrong. Adams notes in his study of problem solving (1974):

"Much thinking went into the mechanical design of various types of prototype tomato pickers before someone decided that the real problem was not in optimizing these designs but rather in the susceptibility of the tomatoes to damage during picking. Part of the answer was a new strain of tomatoes with tougher skins."

From Topic to Aim: The Goal of Your Document

A document is inevitably a slice of your work, one that needs its own structure. Before you begin drafting, therefore, you need to develop an aim for your document. An aim is the reason for writing the document, a specific goal. When you convert a research topic to a document aim, you convert a category to an operation. You propose to do something for someone, namely your audience. Faced with a scanner malfunction aboard Landsat 4, for example, you might aim to demonstrate that Landsat 4 malfunctioned because the X-disk in the scanner jammed against the assembly mount. Or your aim might be to describe a detailed protocol for remotely manipulating the motor of the scanner assembly.

Defining your aim means asking yourself, Why am I writing this document? Usually, your answer to this question tells you that you want to describe physical objects and processes, narrate developments, analyze your topic, or persuade your readers of something. Even highly specialized topics can be approached with different aims, as Figure 2.6 shows.

As you define your problem, identify your aim, and home in on your audience, you create the framework for your writing. One way of pulling all these considerations together is to write a paragraph or two that reflect your thinking. This statement of aim defines your audience, narrows your topic, and focuses on a specific argument. Consider the following project:

In 3 months of work on a control algorithm for remotely adjusting space vehicle direction and speed, two design engineers gather a dozen memoranda and design documents on guidance control strategies. They maintain two notebooks with observations and calculations treating space vehicle design and a new selection algorithm for firing space vehicle jets. They fill another notebook with the minutes of meetings at their home research organization and the contracting government organization. They also develop several large digital files of computer simulations, engineering drawings, rough schematics, tables, and graphs. The research, carried out at a national aerospace laboratory, must be condensed into a report 30 to 40 pages long, including flow charts and the recommended selection algorithm.

For these design engineers, a statement of aim expresses their intentions. Like any statement of aim, theirs should answer three questions:

1. *What is the primary goal?* Try to state your aim in simple operational language that implies action. If the problem is controlling the transla-

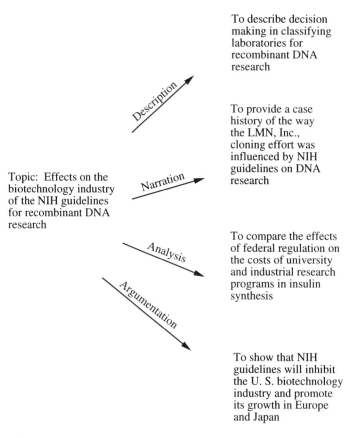

Figure 2.6
By answering the question, Why am I writing this document?, you convert your topic, by means of the above prose strategies, to a specific aim. Each of the aims accurately represents this author's main topic.

tional velocity and orientation of the vehicle, a statement of aim might assert that controlling the translational velocity and orientation of the space vehicle can be viewed as an action—that of initiating firing times for a group of jet thrusters.

2. *What problem is being addressed?* Examine the problem by establishing the situation and conflict you are addressing. Each problem consists of several parts or key variables. By identifying what they are, you establish your perspective on the problem. You also tell your readers just how you are going to treat your material.

3. *What are the main aims of your document?* By stating your objectives in the document, you provide the kernel of your argument. Keep the objectives simple but specific enough for your reader to grasp your method of solving the problem. The objectives determined by our design engineers, for example, could argue that the solution is to adjust jet-firing times by minimizing a quadratic function of the errors in the jet-firing times. Their statement of aim might conclude by noting that the report develops a selection algorithm for remotely firing the jet thrusters.

Aims Imply Audience

The key to drafting a statement of aim is to keep it simple and operational. You strip away most of the qualifying detail to arrive at your central goal, the problem addressed, and your specific objectives. The process, although always difficult, forces you to come to terms with the priorities for your work. It is a process of clarification.

Readers read to solve their own problems. If you state your aim clearly, your potential readers may make informed choices about whether or not to read your work. If you stick to your aim throughout your document, a reader who shares your aim will keep reading. Always ask yourself, What are my intended readers going to do or know after reading my work? You will want to return to this question many times as you draft your document.

3

Organizing and Drafting Documents

You have a problem to address. You've spent weeks on the solution. Your colleagues agree that you're on the right track. You've thought about your audience. You begin to see on paper what exactly you can— and cannot—claim. Your discoveries are about to assume their soon-to-be-transmitted shapes.

Your impulse may now be to sit at your word processor and write your document, from the first page to the last. Chances are that won't work. You now need to think about organization. You will need an outline. Outlining is a powerful means of analysis and synthesis, a tool that helps you develop your prose strategy.

There is no standard way to outline. All outlining is a process of trial and error. Some people work with crude scratch outlines. Others use formal patterns. Still others use templates from word-processing packages, which can help organize material. No program, however, is a substitute for logical thinking. You can't outline merely by following a formula.

Outlining as Organization

The process of outlining partitions your document. You divide your materials under topics, sequence the topics, and then further subdivide them into subordinate ideas. As you arrange topics, you also mold a structure of key points that shapes your work. Outlining effectively isolates and sequences the categories of interest. As you explore the

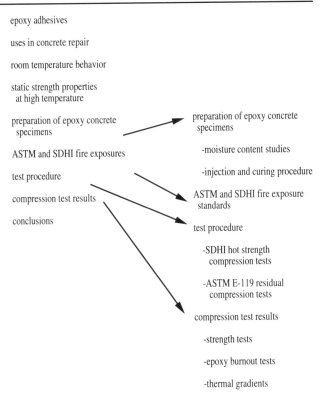

Figure 3.1
Outlining as a process of partitioning your subject by listing topics. Such topic development helps you set priorities and think about connections between topics.

relationships among these categories, you define the limits and emphasis of your document.

Think of outlining as a stage of the writing process. Doing it well means keeping in mind the following:

• *Use outlining to isolate topics.* As you name some topics and dispense with others, you give focus to your document. To develop the parts of an outline, you need to identify keywords that define your categories and reflect your aims. As you list these words and phrases, you can also consider subcategories, as Figure 3.1 shows.

• *Explore the logic of your source material.* This exploratory process eventually leads you to connect the parts of your document logically.

Aim: To argue that stresses generated by the reinflation of the Owen Valley
magma reservoir caused six magnitude 5 earthquakes in July 1983

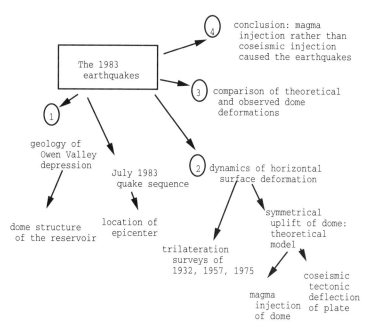

Figure 3.2
An informal scratch outline. Here, we see a series of topics and claims in four
areas that correspond roughly to the sequence of problem, methods, data, evaluation, and conclusion.

You begin to see how to arrange your topics into larger patterns. Scratch
outlining, as shown in Figure 3.2, may reveal a conceptual structure for
your material. Often, however, you have to reconcile this conceptual
logic with the standard formats required by journals and funding
agencies. These formats, some of which are shown in Figure 3.3, are
conventions that assist readers by organizing material in predictable
ways.

• *Integrate a general document design with your specific material.* In
designing a document, try to integrate a general format with the specifics
of your project topics (see Figure 3.4). Your format may be a standard
requirement, or it may be your own design or a series of categories, such
as the case study format of Figure 3.4. As you structure the document,
keep converting the general structure to the more detailed topic outline.

Journal Article	**Formal Report**
[Title, abstract]	[Title page, executive summary, table of contents, list of figures list of tables, nomenclature]
Introduction	Introduction
Theoretical development	Problem
Experimental section	Aim
Materials	Scope
Apparatus	Methods
Procedure	Results
Results	Discussion
Discussion	Conclusion
Conclusion	Recommendations
Acknowledgments	References
References	Appendixes

Design Report	**Memorandum, Lab Report**
[Title, abstract, table of contents, list of figures, list of tables, nomenclature]	Salutation [To:, From:, Subject:, Date:]
Introduction	Problem and background
Design theory and parameters	Discussion and conclusions
System A--introduction	Experimental procedure
Subsystem 1	
Subsystem 2	
System B--introduction	
Subsystem 1	
Subsystem 2	
System assembly	
System performance	
Evaluation	

Figure 3.3

Some typical document structures. Such formats are tools that serve to reduce material into predictable patterns for readers. These formats also summon up logical structures appropriate to specific kinds of subjects, methods, and data.

Identification of Contamination in Electro-deposited Contacts

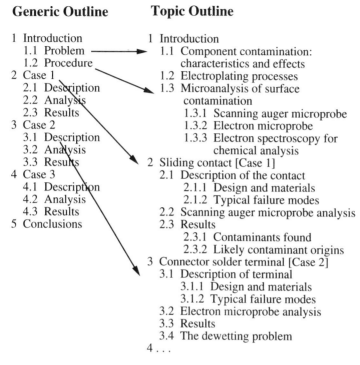

Generic Outline

1 Introduction
  1.1 Problem
  1.2 Procedure
2 Case 1
  2.1 Description
  2.2 Analysis
  2.3 Results
3 Case 2
  3.1 Description
  3.2 Analysis
  3.3 Results
4 Case 3
  4.1 Description
  4.2 Analysis
  4.3 Results
5 Conclusions

Topic Outline

1 Introduction
  1.1 Component contamination:
     characteristics and effects
  1.2 Electroplating processes
  1.3 Microanalysis of surface
     contamination
    1.3.1 Scanning auger microprobe
    1.3.2 Electron microprobe
    1.3.3 Electron spectroscopy for
       chemical analysis
2 Sliding contact [Case 1]
  2.1 Description of the contact
    2.1.1 Design and materials
    2.1.2 Typical failure modes
  2.2 Scanning auger microprobe analysis
  2.3 Results
    2.3.1 Contaminants found
    2.3.2 Likely contaminant origins
3 Connector solder terminal [Case 2]
  3.1 Description of terminal
    3.1.1 Design and materials
    3.1.2 Typical failure modes
  3.2 Electron microprobe analysis
  3.3 Results
  3.4 The dewetting problem
4 . . .

Figure 3.4
Moving from a general to a topically specific outline. Outlines effectively reduce data by partitioning material and creating subject focus and scope. They also show preliminary sequence and logical organization. In addition, they serve as writing and revising aids, as well as furnishing section headings for document design.

• *Adjust the scope and sequence of your source material to reflect your stated aims.* Outlining is a means of data reduction. As you arrange information, eliminate what you do not need. Two common approaches to sequencing information, as seen in Figure 3.5, are logical order and order of importance. Arrangement in logical order is more likely to appeal to experts, whose interests are conceptual; following the order of importance is more likely to appeal to managerial and administrative audiences concerned with costs, personnel, and schedules.

• *Get feedback from colleagues and supervisors.* Circulate your working outline for discussion. One important use of outlining is to promote

Topic: Evaluation of Soil Contamination at Site XYZ outside Building 14N-304

Logical order ⟶ **Order of importance**

Logical order	Order of importance
Executive summary	Executive summary
1 Introduction	1 Introduction
1.1 Purpose of the evaluation	1.1 Purpose of the evaluation
1.2 Team makeup and schedule	1.2 Scope of the visit
1.3 Objectives of the visit	1.3 Team makeup
1.3.1 Event evaluation	2 Summary of findings
1.3.2 Scope of the visit	3 Recommendations
2 Chronology of the occurrence	3.1 General safety measures
2.1 Background	3.2 Specific site actions
2.1.1 Facility description	4 Causal analysis
2.1.2 Records evaluation	4.1 Direct cause: decontamination
2.1.3 Assessment approach	procedure
2.2 Event chronology	4.2 Root and contributing causes
3 Causal analysis	Appendix A. Assessment methods
3.1 Root cause	Appendix B. Facility description
3.1.1 Drain line failure	Appendix C. Event chronology
3.1.2 Design of sewers	
3.1.3 Detection system	
failure	
3.2 Direct cause: Decontamination	
procedure	
3.3 Contributing causes	
3.3.1 Communications	
3.3.2 Records management	
3.3.3 Regulatory compliance	
4 Recommendations	
4.1 General safety measures	
4.2 Specific site actions	
5 Summary	

Figure 3.5
Logical order vs. order of importance. Arranging material according to its logical order is likely to appeal to other specialists. Arranging material to reflect its order of importance is likely to appeal to managers and administrators.

consensus on goals, coverage, and strategy for documents, whether proposals, articles, or theses. At the outlining stage, the suggestions of a collaborator or supervisor may save you considerable time later in the writing process.

Drafting the Document

Drafting is never the same for any two writers. Methodical writers work from an outline, point by point. Intuitive writers may write entire sections at a time, barely glancing over their outlines. Your outline is a map that should keep you on course and remind you of your aim. You may alter it frequently as you write. You may enter it into your word processor and expand it to create a crude draft. For example, you might develop an outline entry such as "moisture in concrete samples" through an assertion:

Because moisture is never entirely absent from concrete, simulated fire exposure studies should factor in a heat absorption capacity for an amount of water 2.5% of the total sample weight.

With such claim statements, you expand your main arguments outward from the outlining stage.

Tools and Tips for a First Draft

Word-processing programs are effective tools for merging the outlining and drafting processes. You can develop an initial outline, start writing individual sections of it, and then rework material that needs further development. Your outline can be fleshed out and your draft revised in whatever order you feel comfortable with. Some writers type the draft into a word processor; others write it in longhand, have it word-processed, revise on a printout, and have it word-processed again.

Here are some tips for drafting:

• *Review your aim.* Try to keep your main writing goals in view and avoid digressing.
• *Set writing goals.* Writers write drafts most successfully in stages. Set an objective for a four- or five-page section and write the section at one sitting, if possible.
• *Maintain momentum.* Keep writing. Don't insist on achieving finished copy. Don't worry about where you start. You can begin with the concluding section and end with the introduction; that way, your conclusions will be focused on your introductory claims. By writing out of document sequence, you can build up writing momentum for more difficult sections.
• *Revise rigorously in hard copy.* Computing encourages wordiness. If you print out the text, you can get a sense of the whole, and you can jump around quickly as you edit for organization and consistency.
• *Expect to go through several draft and printout cycles.* Use the revision capabilities of computing to review and revise your draft. Don't put all your time into your first draft.

The very first draft, usually called the rough draft, is something you generally don't show to anyone. Few people other than you will be able to read it intelligently. A rough draft is useful, although sloppy, because it allows you to organize your document and work out your main arguments. Your next draft may still be crude, but it may be ready for the attention of others.

Cycling a draft can help you immensely if it brings constructive criticism. Cycling is a requirement in many organizations, but you should follow one firm rule: Don't submit crude drafts to colleagues or supervisors. Crude drafts often get treated as final drafts. A crude draft may be nearly complete, but you don't really have a finished draft until you have the following:

• An introduction, middle development, and conclusion, so that coverage and analysis can be examined
• Coherent, grammatical language
• Accurate spelling
• A series of clear, if still rough, graphics
• Clean copy, with subject headings, subheadings, and standard margins

Be sure you've reached this stage before you begin to circulate your copy to those who will judge your work. Readers soon forget the distinction between rough and final copy, and they will inevitably associate your performance with what they see first. Their opinions of your efforts may then be lasting.

Argument

An argument supports a claim with a convincing set of reasons. You can expect to make arguments in nearly all documents, whether technical reports, refereed journal articles, or memoranda. A writer's aim is rarely just to inform. You need to identify problems, make claims, and defend them convincingly. Generally, the facts, no matter how effectively you analyze them, will not unequivocally support your claim. Writers need to interpret the facts to show that they mean what the writer thinks they mean.

Facts rarely speak for themselves. Readers need to understand the context of the facts. They must be convinced that the facts are accurate and effectively used. Hence, readers need to know exactly what your claim is, what problem you think you are addressing, what you think are the issues of the argument, how you got your evidence, and what you think your evidence means. Your argument can be strengthened or weakened at any of these steps.

If your readers think you have a false problem, they will doubt your judgment. If they don't think you have developed an effective approach to solving the problem, they will think you are missing the point. If they think that your evidence is weak, they will doubt the rigor of your effort. If they don't like the way you used your evidence, they will question your reasoning. And if you manage to satisfy your readers on all these points, they may still argue that you came to the wrong conclusions.

One way to construct arguments is by linking together:

1. A problem or situation to be remedied
2. A claim or thesis that resolves the problem
3. Background issues that give the particulars of the problem and establish criteria for solving it
4. Evidence that supports the claim or thesis that applies the criteria
5. A discussion in which the evidence is weighed and shown to support the claim

This general structure (Figure 3.6) is used repeatedly, quite independently of a writer's field of specialization, profession, or job definition. These elements of argument do not always have to be explicit, although they usually should be. Moreover, the order and extent of their development will depend on the kind of audience you are addressing.

In the short memorandum shown in Figure 3.6, we see these elements in an argumentative sequence. We see a problem that has a commercial and a technical aspect. The problem definition is narrow, but the argument, in this instance, is merely that a formal project should be undertaken to establish the effectiveness of the proposed solution.

A problem, once divided into so many issues, now demands certain kinds of evidence. Hence, an argument should be considered a system of prose elements, a conceptual plan or prose strategy for making the most of your evidence. In Figure 3.6, the authors have limited the terms of the problem-solving effort to the laboratory checking of thermal stability of the proposed scale inhibitor. They have established criteria for the test, even though in two instances the performance of the $CaCO_3$ inhibitor did not reach the upper limit of the test criterion.

Memorandum

TO: M. White, Manager, Support Products Division
FROM: J. Kline, CDT
SUBJECT: Product Developments--New Scale Inhibitor
DATE: July 22, 1994

Scale Inhibitors--Need for High-Temperature Performance
    Condex, Inc. currently markets no effective scale inhibitor for controlling calcium
carbonate deposits in multi-stage flash evaporators operating in the temperature range of
240-280 F. We may need this product in order to maintain a strong competitive
international standing as production contractors for water distillation plants.
    Some informal initial laboratory tests indicate that Condex's DPT-62 might be an
excellent potential scale inhibitor that would perform well at temperatures from 230-285 F.

Problem defined both technically and commercially

Stability Requirements in Relation to DPT-62
    The main technical problem in developing a new scale inhibitor is thermal stability.
Although there are several candidates among the polyacrylates, few of these are stable over
temperatures of 240 F. Yet, new flash-evaporator technology will routinely operate at
temperatures of 260 to 280 F within 5 years. Our standard scale inhibitor, H_2SO_4, is too
caustic at temperatures higher than 240 F. A good commercial $CaCO_3$ inhibitor would
        1. Prevent the formation of scale at temperatures between 240-
        280 F
        2. Have thermal, hydrolytic, and oxidative stability at temperatures
        between 240-280 F
        3. Conform to EPA standards for toxicity in water-treatment
        environments.

    Lab tests, using a scale adherence simulator (SAS), show that DPT-62
        1. Inhibits $CaCO_3$ scale at temperatures between 220-280 F
        2. Has thermal stability at 240-285 F
        3. Has hydrolytic stability at 240-275 F
        4. Has oxidative stability at temperatures between 240-270 F
        5. Is effective in doses that will not exceed EPA standards

Evidence that addresses the problem and supports the conclusion

Recommendations
    Laboratory testing, carried out within CDT over the past two months, has established the
effectiveness of DPT-62 under most of the conditions set. I recommend
        1. Initiating a full field-test at the Woods Hole, MA site
        2. Building a new laboratory testing unit (SAS) for further scale
        inhibition studies on magnesium hydroxide
        3. Initiating comparative performance tests on the competition's
        inhibitor, T-XYZ.

Conclusion consistent with the aim of the document

Figure 3.6
Argumentative structure, as developed in an internal memorandum. This brief
recommendation report displays the essential problem identification, followed by
the criteria, evidence, and conclusion.

Evidence can be judged only in the context of such criteria, and it may not always perfectly fit the conclusions. But evidence is always interpreted in some larger context of possibilities and needs.

From Drafting to Reviewing to Revision

Eventually, your draft will be ready for reviewing. Before your work is in final form, you will want the appraisal of your colleagues and probably your supervisor as well. The time it takes to circulate your document so that others have a chance to read it and make suggestions is your time for a breather. Many writers put their work on the shelf and move on to something else during this period. When they come back to their work to incorporate comments and revise, they have a fresh perspective.

4

Revising for Organization and Style

A team of biologists hurries to finish a large proposal for an important funding agency. Pressed for time, team members suspend their laboratory work while they draft their separate sections of the proposal. With 24 hours to go before the deadline, the team meets to read the complete document, revise it, and compile the sections.

Unfortunately for team members, their very worthy project is unlikely to be funded. Their aims may be well stated and their plans well developed, but they've made an all-too-common mistake: They've left too little time for revision.

Revising is part of writing. To achieve the high level of organization and clarity necessary for your writing to succeed, you need to learn to revise your work. When writing goes out for delivery, the writer no longer controls the message. Your written thoughts are available for unlimited close analysis, "operating" on your behalf in your absence. Good writing can reduce great volumes of descriptive and analytical detail to refined, closely reasoned arguments. The finished document must be clear, accurate, and consistent. You won't be personally present to fill in gaps or correct errors and inconsistencies.

When revising, writers attempt to improve the substance, organization, and clarity of their prose. Approach this phase with an open mind. Here are some things to consider:

• *Organization.* Can you reorganize the draft in ways that will make its structure more closely reflect your goals and the needs and interests of your readers?

• *Logic.* Will adding, deleting, or rearranging material noticeably strengthen its logic and coherence for readers?

• *Focus.* Can you rewrite paragraphs to focus on the subject matter more effectively for your readers?

• *Accuracy.* Is the prose detail sufficiently accurate and complete to support your claims?

• *Clarity.* Can you make the prose more coherent and accessible for readers?

• *Economy.* Can you condense or eliminate wordy paragraphs or sentences?

Writers need to gain perspective on their own work. You get some distance by setting your written outlines and drafts aside for a time and by asking for input from colleagues and supervisors. If you have been immersed in your project for several weeks, you will often miss obvious gaps in your writing. You tend to read in missing material. You also tend to understand your written work in the order you have fixed in your mind. Readers' questions and criticisms will let you know whether you're communicating with your audience. So don't be defensive; you don't have to accept every recommendation.

Organization First, Then Style

The distinction between organization and style is a matter of emphasis, because effective organization is good style. Organizational revision encompasses a larger scale of activities. You add, delete, or alter content to improve its logic and focus. In stylistic revision, you rework individual sentences, rephrasing for greater impact on your audience. Your goal at the earliest revising stage is to make the draft reflect more accurately your aim and writing strategy. Concerns about clarity, word choice, and economy are later, stylistic problems.

As Figure 4.1 shows, manuscripts are most effectively revised and prepared in a certain order. Once you have satisfied yourself about your coverage and organization, you try to develop clearer paragraphs and sentences, effective transitions between sections and paragraphs, and accurate use of words and detail. At the later stages of preparation, the manuscript needs to be edited for grammar, punctuation, and mechanics,

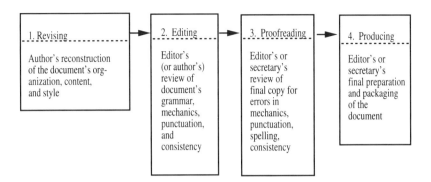

Figure 4.1
Stages of revising, editing, proofreading, and producing documents. Note that various individuals may play roles at the different stages. For some documents, these stages may not be distinct.

as well as proofread for various inconsistencies and omissions. Finally, it must be produced for distribution. It's good to know how to do all this yourself, even though you may have secretarial or editorial assistance in the later stages of the process.

When revising, first read the draft and circle material that you may want to relocate or delete. Jot notes in the margins, but don't start re-arranging or cutting sections until you have reviewed the whole manu-script. Once you have familiarized yourself with the draft, then study your original outline to see if your manuscript meets your expectations.

Here are some questions to ask yourself:

• *What results do you want from circulating the document?* What claims are you making? What criteria have you set for achieving your claims? What do you want your readers to do after reading the document? Does your document make this response clear?

• *Will your document work for the audience you are trying to reach?* Is the organization appropriate for your readers? Have you thought care-fully about material they need in your document to reach their goals? Is the level of your material (including your graphics) appropriate for the audience?

• *Are the aims of your document clear?* Does your introduction contain your document's aim and problem statement? Have you stated the prob-lem clearly? Keep in mind that your problem statement may not become clear until you have worked it over in draft and can see how the criteria and evidence have lined up. Don't elaborate in such detail that the problem becomes obscured by the detail.

If you decide that you want to reorganize your document, consider revising your outline to establish a new plan. Then cut and paste sections in the new sequence, using a photocopy or duplicate word-processing file, so that you can retreat to the original revision if the rearranged draft doesn't come out right. Remember that once you have cut and rearranged the draft, you will probably have to write new transitions.

Organization Is Content

Much revision is devoted to reorganizing and developing material. For example, your theoretical section may belong not after your introduction but after the discussion of your experimental results. You may have given an account of your methodology in the results section of your report and want to shift this preliminary material back to its own section. You may want to give a more detailed account of the problem you are addressing. You may have buried important recommendations in back sections, where the managerial reader will have trouble finding them. These and similar problems would require you to rearrange considerable amounts of material.

You often need to develop material that has not been smoothly or logically presented in the draft. Revising is partly a matter of rearranging sections to (1) tailor the detail to specific aims and audiences, (2) improve logical development of the material, (3) add or delete detail for the sake of argument, or (4) highlight important material. You can often improve the focus and logic of your writing if you think about ways to organize more effectively.

Reorganizing for Greater Audience Impact
To recorganize, work with an outline, which you can use as a plan for rearranging or deleting. Note, for example, the two outlines for an in-house administrative report shown in Figure 4.2. The authors were recommending office automation and approached the question logically, sorting out and discussing the significant factors of the technology. Organizing the material by discussing each type of system helped the

Office Automation at LMN, Inc.: A Two-Year Plan

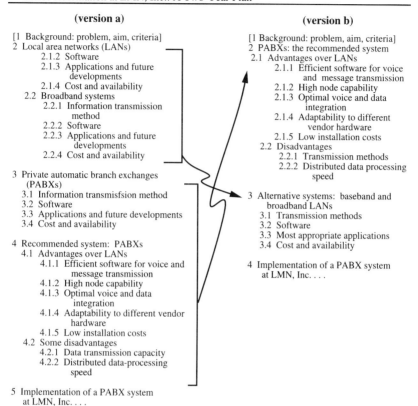

(version a)

[1 Background: problem, aim, criteria]
2 Local area networks (LANs)
  2.1.2 Software
  2.1.3 Applications and future
      developments
  2.1.4 Cost and availability
  2.2 Broadband systems
    2.2.1 Information transmission
        method
    2.2.2 Software
    2.2.3 Applications and future
        developments
    2.2.4 Cost and availability

3 Private automatic branch exchanges
  (PABXs)
  3.1 Information transmisfsion method
  3.2 Software
  3.3 Applications and future developments
  3.4 Cost and availability

4 Recommended system: PABXs
  4.1 Advantages over LANs
    4.1.1 Efficient software for voice and
        message transmission
    4.1.2 High node capability
    4.1.3 Optimal voice and data
        integration
    4.1.4 Adaptability to different vendor
        hardware
    4.1.5 Low installation costs
  4.2 Some disadvantages
    4.2.1 Data transmission capacity
    4.2.2 Distributed data-processing
        speed

5 Implementation of a PABX system
  at LMN, Inc. . . . .

(version b)

[1 Background: problem, aim, criteria]
2 PABXs: the recommended system
  2.1 Advantages over LANs
    2.1.1 Efficient software for voice
        and message transmission
    2.1.2 High node capability
    2.1.3 Optimal voice and data
        integration
    2.1.4 Adaptability to different
        vendor hardware
    2.1.5 Low installation costs
  2.2 Disadvantages
    2.2.1 Transmission methods
    2.2.2 Distributed data processing
        speed

3 Alternative systems: baseband and
  broadband LANs
  3.1 Transmission methods
  3.2 Software
  3.3 Most appropriate applications
  3.4 Cost and availability

4 Implementation of a PABX system
  at LMN, Inc. . . . .

Figure 4.2
Reorganizing content for audience impact. In this example, the material is rearranged from logical order (version a) to order of importance (version b) as a way of accommodating decision making. Note that the shift not only reorders but also condenses the original material.

authors learn the issues but did not meet the needs of the audience, who had to take action.

In the revised organization, the writers reflect a pragmatic order of importance. To gain acceptance for the recommended system, they first discuss the advantages of private automatic branch exchanges (PABXs) of local area networks (LANs). Having firmly established the reasons for their choice of system, they survey the alternatives. Note that considerable material has been condensed in the second version to deemphasize the systems not recommended. This kind of large-scale revamping and condensation of draft documents may be worked out first in an outline. Once you have the basic structure determined, you can review and revise with other goals in mind.

Make Fragmentary Prose Coherent   Be sure you convert your outline to prose. Sometimes, as in Figure 4.3, your draft statement will be almost in outline form. The result is an awkward and poorly focused passage. Concentrate now on making sure that you have paragraphs; you can revise them in the next stage. A paragraph needs a clear topic sentence, and it might refer the reader to a graphic. The remaining detail should then be subordinated. These measures will develop the logical flow of your material.

Expand Prose to Strengthen Arguments   Draft material often needs to be expanded. In Figure 4.4, for example, the weak version is unfocused, partly because the prose is too compressed and vague. The material needs to be sorted and developed with more care. Although the weak version is comprehensible, phrases like "high-cost industry" and "small profits involved" are weak generalizations. Moreover, the second sentence attempts to make so many different points that it sacrifices specific detail. In the improved version, the initial topic sentence is more clearly focused. Rather than including all the information in two rambling sentences, the author now develops the same concepts through the logical progression of four sentences, with information linked through words such as "because," "moreover," and "however."

Arrange Material to Frame Discussions More Effectively   The management of detail is a major problem in nearly all scientific and technical

Weak version

3 Experimental procedure

3.1 Apparatus

3.1.1 <u>Chopper wheel</u>:
Diameter--approx 1.2 cm
Material--cadmium, slit width approx. 1 mm.
No. of slits--6.

It's purpose is to chop the neutron beam into
clusters of particles, whose energy distribution will
correspond to the Maxwell-Boltzmann distribution.

3.1.2 <u>Motor (low power AC)</u>. Its purpose is to
rotate the chopper wheel.

Improved version

3 Experimental procedure

    3.1 <u>Apparatus</u>. The main apparatus used for
neutron counting is a chopper wheel, shown in
Figure 3, that chops the neutron beam into clusters
of particles. The energy distributions of the clusters
should correspond to the Maxwell-Boltzmann
distribution. The wheel, as shown, is a cadmium
disk, approximately 1.2 cm in diameter, with 6 slits,
each approximately 1 mm in width. The wheel is
rotated by a low-power AC motor.

Figure 4.3
Expanding fragmented prose to improve logical presentation. This kind of revision is helpful when the writer has not got very far past the outline stage. Note
that a graphic (not shown) is added in the improved version.

Draft version (2 sentences)

The current photovoltaic industry is a high-cost industry, whose production of solar cells is labor-intensive and therefore costly, and this limitation makes investors reluctant to invest because of the small profits involved. According to recent studies (Clifford, 1986), this is a problem for manufacturers who must decide whether to further intensify their labor for possible short-term profits or to seek government support and to automate.

Revised version (4 sentences)

The high prices associated with limited production of solar cells have prevented the photovoltaic industry from attracting investment capital needed to automate manufacturing lines. Moreover, a recent study indicates that large companies in the United States will be unwilling to invest in photovoltaic array production until an annual market of $50 to $100 million is certain (Clifford, 1986). In order to meet the increased near-term demand for solar cells, manufacturers may find it cheaper and easier simply to add workers to the manufacturing line, rather than to invest in new production facilities. This choice, however, is complicated by recent government interest in making large-scale purchases of solar cells in order to stimulate price reductions.

Figure 4.4
Expansion of a draft to improve the coherence of the argument. In the improved version, the writer develops the material more gradually in four rather than two sentences. This expansion is meant to help the reader sort out the logic of the argument. A great deal of revison is normally devoted to this kind of prose reconstruction.

prose. Not only are there immense quantities of physical detail, but the detail is also often repetitive. Organize information so that the details don't obscure your aims. In Figure 4.5, the weak version makes a poor start with an extensive listing of minor details. When you place such routine enumerations in the most prominent and strategic place, you cloud the purpose of the section. In the improved version, the italicized detail has been shifted into Table 3. This sorting of detail enables the author to concentrate on the experimental results.

Delete Detail that Does Not Advance the Discussion   Be prepared to cut the many digressions that typically evolve during drafting. Does the detail truly support your argument? You might find that some of your information can be deleted or moved to another part of your document.

Highlighting Material that Traces Your Argument

Highlighting and subordinating enhance your organization. Several common strategies help improve the focus and clarity of your draft.

• *Repeat major ideas or themes.* Repetition, used sparingly, can help maintain the focus of your work. You can restate a primary theme of your introduction at the beginning of one or two middle sections of the document and again in the conclusion. Such a restatement, as Figure 4.6a shows, reestablishes the document's main argument.

• *Use headings and subheadings.* Headings and subheadings are conventional but often underused aids. By signaling the next topic (see Figure 4.6b), they help the reader assess the logic and coherence of the material. They announce shifts in the discussion, supply transitions between sections, and together provide an outline. Your system of headings and subheadings should draw directly from your original rough outline.

• *Use graphics.* Graphics are the most potent techniques for emphasis. They may be used to highlight important subject matter and to delineate concepts too complex to treat in prose. The simple schematic can clarify complex relationships, emphasize key concepts, and communicate more rapidly than prose.

• *Pay attention to parallelism.* Itemizing and enumerating are common techniques for creating emphasis. These formatting devices show parallelism between concepts. They clarify sequence by drawing attention to the series itself. In Figure 4.7, enumerations draw attention to a series of steps. The same principle can be used to list recommendations or highlight other elements.

Weak version

Experimental Results

All the tests were carried out using a pressure of 118 atm. The surface temperatures of the pipe at the time of crack initiation were $+12^{o}C$ (Test A1), $+3^{o}C$ (Test A2), $+8^{o}C$ (Test A3), $+6^{o}C$ (test B1), $+12^{o}C$ (Test B2). In all tests, the crack propagated axially along the top of the test pipe, then turned abruptly into a circumferential loop, after which it was arrested. This suggests that the crack was arrested by the notch ductility of the pipe itself.

Crack Propagation and Its Velocity

A rough sketch of the crack path and the changes in crack velocity are shown in Figure 2. *In Test A1, the crack was arrested in the N2 pipe at 188J Cv on the North side (QT pipes) of the crack initiation pipe (CIP); on the South side (CR pipes), the crack was arrested in the S3 pipe at 202 J Cv. In test A2, the crack was arrested in the N3 and S3 pipes at 275 J Cv, respectively. In test A3, the crack was arrested in the N3 and S3 pipes at 202 J and 196 J; in Test B1, it was arrested in the N2 and S2 pipes at 156 J and 126 J; and in Test B2, it was arrested in the N2 and S2 pipes at 129 J and 145 J.*

In the A-series tests, just after crack initiation, the crack velocity reached more than 300 m/s in the CIP, which had a relatively low Cv of 50 J. Then after the crack entered the test pipes, . . .

Improved version

Experimental Results

Crack Propagation and Its Velocity

In all tests, the crack propagated axially along the top of the test pipe, then turned abruptly into a circumferential loop, after which it was arrested. A rough sketch of the crack path and the changes in crack velocity are given in Figure 2, *and crack arrest locations and their respective Charpy energies (Cv) are identified in Table 3.* The characteristic crack pattern suggests that the crack was arrested by the notch ductility of the pipe itself.

In the A-series tests, just after crack initiation, the crack velocity reached more than 300 m/s in the crack initiation pipe (CIP), which had a relatively low Cv of 50 J. Then, after the crack entered the test pipes, . . .

Table 3. Crack arrest locations in the test pipes. QT (quenched and tempered) pipes and CR (controlled rolling) pipes are respectively on the N and S sides of the crack initiation pipe.

		Crack Arrest			
		QT		CR	
	Surface				
Test [a]	T(^{o}C)	Pipe No.	Cv(J)	Pipe No.	Cv(J)
A1	12	N2	188	S3	202
A2	3	N3	275	S3	196
A3	8	N3	212	S3	209
B1	6	N2	156	S2	126
B2	12	N2	129	S2	145

[a] All tests carried out at 118 atm.

Revising for Style

Good style means economy and clarity. It does not seek strict uniformity in prose but does demand vigorous expression. Clarity of style applies to paragraphs, sentences, phrases, and words. Clarity governs these choices throughout the document.

• *Documents.*   Clarity at the document level is generally achieved by document design and apparatus, including abstracts, tables of contents, executive summaries, headings, numbering systems, and white space. All these elements help readers move around in the material.

• *Paragraphs.*   At the paragraph level, clarity refers to coherence, normally achieved by ordering sentences in recognizable patterns of description, narrative, or analysis.

• *Sentences.*   Clear sentences generally develop from balanced elements, without wordiness, with specific detail, and with subject and verb in an unambiguous relationship with each other.

• *Words.*   Clarity of word choice means appropriate and accurate word usage.

The Paragraph

Paragraphs should fit naturally within the whole and manifest an internal order of their own. Tightly constructed, they allow you to advance your work in stages—around key ideas or topics (see Figure 4.8).

A paragraph should provide an effective transition, clear topical focus, and coherent development. Figure 4.8 exemplifies some typical paragraph elements. Commonly, each paragraph is linked with the preceding paragraph, makes some kind of generalization or claim in a topic sentence, and then develops this claim in a series of supporting sentences.

Figure 4.5
Rearranging detail for clarity and focus. The weak version buries the main results in a sea of routine detail. The repetitive numerical detail of the weak version is more effectively placed in Table 3 in the improved version. (Adapted from Sugie et al. 1982)

(a) Repetition

IV. Nematic behavior in other systems

A. Soap solutions. In all of the above systems, we have been dealing with long-range operational ordering (LROO) in liquids composed of small, rigid, anisotropic molecules. *We have seen that LROO (e.g. isotropic-to-nematic phase transition) occurs because at low temperature and high density the decrease in rotational, "mixing" entropy is offset by a lowering of the potential energy and an increase in the translational, "packing" entropy. The details of the isotropic-nematic phase transition depend on the specific features of the attractions and repulsions between pairs of particles, as well as on the extent of their molecular biaxiality, and so on. But the key qualitative feature, the basic driving force for the LROO, is simply the anisotropy of the interacting particles.* Thus it is not surprising that isotropic-nematic phase transitions have recently been observed in systems that have no direct physical connection with the usual liquid crystal circumstances.

Consider the first case of aqueous solutions of amphiphilic (soap) molecules. A typical example of sodium dodecyl sulfate (SDS), shown in Figure 6 . . .

(b) Headings and Subheadings

. . . Now that we have established a basis for understanding how a telecommunications system performs its functions, we shall describe how such a system was implemented on the Voyager.

Telecommunications Design on the Voyager Spacecraft
The design of any telecommunications system begins with an examination of the mission requirements for telemetry, command, and radio metrics. For Voyager, the telemetry data requirements . . .

Figure 4.6
Two ways to emphasize material. In (a), the material in italics repeats an earlier overview and places the new discussion in the context of the paper's main theme (Gelbart 1982). In (b), the author effectively moves from the general discussion of telecommunications systems to the specific instance of the Voyager spacecraft (Edelson et al. 1979).

Weak version

A paradigm for stochastic robot feedback follows several steps. First, we put the peg into the hole (or next hole), by aiming the robot at the point where the hole is. We then measure, as well as possible, the position of the robot and its offset, with respect to the hole. Next, we estimate the position of the hole and of the robot. This estimation provides the basis for estimating the position of the next hole. Then, we return to the first step. In place of next hole, we could read next item or pallet or next instance of this hole on the next workpiece, and so on.

Kalman filtering provides a natural framework for developing this paradigm. . . .

Improved version

A paradigm for stochastic robot feedback follows several steps:

1. Put the peg into the hole (or next hole), by aiming the robot at the point where the hole is.
2. Measure, as well as possible, the position of the robot and its offset, with respect to the hole.
3. Estimate the position of the hole and of the robot.
4. From this, estimate the position of the next hole.
5. Go to 1.

In place of next hole, we could read next item or pallet or next instance of this hole on the next workpiece, and so on.

Kalman filtering provides a natural framework for developing this paradigm. . . .

Figure 4.7
Visual emphasis through formatting. Enumerating focuses on a series of operational steps that shows the sequence of expected actions. (Adapted from Whitney 1981)

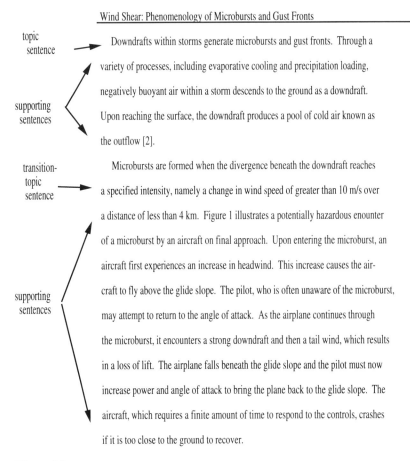

Figure 4.8
Two paragraphs and their elements, developing in conventional patterns from transition and topic sentences to supporting sentences. (Adapted from Merritt et al. 1989)

Diagnosing a Weak Paragraph   Faulty paragraph organization, weak sentence structure, wordiness, inaccurate word choice—all are common problems. For example:

The main operating cost is hydraulic oil replacement. GR-30 oil costs $18.60K per total plant charge, compared to XR-1023, which is $7.80K. Oil life has a major impact on the overall economics. Preliminary economics were based on a 1-year nickel life, while XR-1023 has historically seen a 2-year life in the various units of the plant. The shorter life was taken generally to be due to a lack of experience and the possibility that chlorine in the hydraulic system may accelerate the degradation of GR-30.

In this draft, we see some common problems in paragraph, sentence, and word use. The paragraph is not coherent. The first sentence has no mention of the main subject, for which catalyst replacement is being considered. By not focusing the paragraph, the authors leave their readers to struggle with a series of statements that are not tied together by the cost concerns of the authors. The first three sentences thus do not delineate an idea. Then, in the fourth sentence, the author begins with "preliminary economics" without ever establishing what the term actually means.

Furthermore, some sentences are weak. In the fourth sentence, the subject of the first clause, "preliminary economics," has no clear parallel with the subject of the second clause, "XR-1023." The authors make this problem worse by shifting oil terminology of "GR-30" to "nickel." Word usage is also vague, as phrases like "overall economics" and "preliminary economics" are undefined. Phrases like "the shorter life" and "lack of experience" are not attached to any clear subject.

Making Paragraphs Coherent   In poorly organized paragraphs, an author commonly fails to establish a clear topic sentence. The supporting sentences, left without a clear idea to develop, tend to operate as a group of related but autonomous statements. Readers must construct the argument themselves.

To revise such paragraphs, rewrite the main sentence, so that it clearly establishes the key idea. Once this focusing sentence is in place, you have a concept under which you can line up your supporting material. Consider this revision, which establishes that hydro plant economics depends on hydraulic oil cost and life. Supporting sentences then focus on aspects of oil cost or life, and consistent use of terms helps make key distinctions.

Hydraulic oil cost and life largely determine the economics of the hydro plant, because its main operating cost is oil replacement. GR-30 and XR-1023, the two oils we are considering, cost $18.60K and $7.80K, respectively, per total plant charge. On the basis of 9 months of plant use, we estimate that GR-30 will average a 1-year life per charge in the various units of the plant, whereas XR-1023 has averaged 2 years per charge over the past decade. The shorter life of GR-30 may result from our lack of plant experience with it and from its accelerated degradation by chlorine contamination of the hydraulic system. We are still projecting, then, that XR-1023 will cost less and last longer per plant charge.

Note that to make a topic sentence, the reviser combines the first two sentences of the draft version, giving a clearer overview for the paragraph. The original topic sentence, "The main operating cost is hydraulic oil replacement," is subordinated to a new main clause, originally sentence 3 (slightly altered). This new clause draws out the primary subjects being discussed. The new topic sentence, "Hydraulic oil cost and life largely determine the economics of the hydro plant, because its main operating cost is oil replacement," links two ideas together with the new reference to the hydro plant.

This new topic sentence effectively focuses the paragraph, providing a context for the supporting sentences. The author can now line up the supporting sentences. Note that a pattern of subjects and verbs emphasizes comparison and contrast by systematically drawing the reader's attention to the main points of the comparison.

This paragraph also adds detail for focus and clarity. Much of the new information was implicit in the first draft, but in the revised version, the authors state some key assumptions. In the original paragraph, the hydro plant was nowhere mentioned as the main subject under discussion. In the new version, the authors note that their estimations of hydraulic oil life are based on 9-month and 10-year operating periods. They note further that the chlorine in the hydraulic system is a form of "contamination." Finally, they add the summary statement that the current economic projections assume that XR-1023 is still the more economical oil. Made explicit, these details strengthen the authors' argument.

Using Parallel Headings to Reveal Logical Flow   Technical subject matter is often so dense with terminology and operations that even well-

3.4 *Unresolved Issue Number 4*. The criteria for restarting Facility XYZ have not been met for the water drain capacity of the filter compartment, the stability of the charcoal in the absorber, and the capacity of the absorber.

The above unresolved issue consists of three separate restart criteria. The details of these criteria are as follows. The first criterion is that filter compartment water drains shall be demonstrated to be capable of meeting their design function. The second criterion is that the possible iodine desorption and autoignition that may result from radioactivity-induced heat in the carbon beds shall be considered when determining the adequacy of the charcoal absorbers. Finally, the third criterion is that the absorber section of the XYZ facility shall contain impregnated activated carbon filters demonstrated to remove gaseous iodine from influent. The carbon filters must have an average atmosphere residence time of 0.25 seconds per 2 inches of absorbent bed. The maximum loading capability shall be of 2.5 mg of total iodine per gram of activated carbon, and no more than 50 mg of impregnate per gram of carbon should be used. The radiation stability of the type of carbon used shall have been demonstrated and certified.

The capacity of the water drains (the first criterion) is addressed in RRD-RSE-910003, "Revision to Filter Compartment Drain Capacity," (21 January 1991) . . .

Revised version

3.4 *Unresolved Issue Number 4.*
The criteria for restarting Facility XYZ have not been met for the water drain capacity of the filter compartment, the stability of the charcoal in the absorber, and the capacity of the absorber.

3.4.1 *Criteria for Restarting*
The following criteria must be met before Facility XYZ may resume operations.

> 3.4.1.1 *Capacity of filter compartment drain lines.* The filter compartment water drains must be demonstrated to be large enough to handle the capacity called for in the design.

> 3.4.1.2 *Stability of the absorber's charcoal bed.* The absorber's charcoal bed must be shown to be stable enough to prevent any possible iodine desorption and the autoignition that might result from radioactivity-induced heat in the carbon beds.

> 3.4.1.3 *Use of carbon filters in the absorber section.* The absorber section shall use impregnated activated carbon demonstrated to remove . . .

3.4.2 *Assessments and Conclusions*
The above criteria may be met as follows:

> 3.4.2.1 *Capacity of the filter compartment drain lines.* Guidelines for regulating the capacity of the water drains are addressed in "Revision to Filter Compartment Drain Capacity" (RRD-RSE-910003, 21 Jan 1991), . . .

Figure 4.9
Use subheadings to show subordination and parallelism. (Adapted from Department of Energy 1991)

designed paragraphs are difficult to follow. Headings mark out topical patterns in otherwise opaque prose. In Figure 4.9, for example, the revised version communicates at a glance the essential logic, and this explicit structure in turn enables the reader to get information out of the paragraph more effectively.

The Sentence

You can often solve paragraph problems by correcting sentence-level errors. The variations of sentence order and modification may seem limitless, but sentences do develop in a few basic patterns. Knowing these patterns will help you analyze and improve weak passages. Identifying sentence patterns will also help you vary your sentence structure. Nothing is more monotonous than a series of identically structured sentences.

The three basic sentence patterns are the simple, compound, and complex sentence. The differences among these are the number and relationships of clauses, word groups made up of a subject and a verb, or predicate. The simple sentence is a single clause, which may include various adjectives, adverbs, and phrases as modifiers. The following are simple sentences because both contain a single subject and a single verb:

The phytoplankton are light sensitive. They absorb oxygen rapidly at 60°C.

Simple sentences may be combined, with the help of a conjunction, to make a compound sentence:

The phytoplankton are light sensitive, and they absorb oxygen rapidly at 60°C.

In a complex sentence, the two clauses have a hierarchical relationship, so that the main clause is linked to a dependent clause by a subordinating conjunction:

If they are light sensitive, the phytoplankton absorb oxygen rapidly at 60°C.

Analyze your writing as you revise for style. Do most of your sentences fit just one of these patterns? If they do, look for ways to vary your sentence structure by combining sentences by using different conjunctions. Varied sentence structure, with sentences of different lengths, will make your writing livelier and easier to read.

You can't, of course, solve all grammatical and stylistic errors by varying your sentence structure, but learning to analyze sentences will help you avoid some common problems. Weak sentence structure clouds meaning. Attention to sentence structure should clarify your writing.

Breaking Long Sentences into Manageable Units   Long sentences, often amounting to more than 30 words, are usually too complicated. Determine the main actions of the sentence. Then sort these into two or more shorter sentences.

Weak: In gasoline engines, designers leave a space between the piston and its cylinder that contributes to the exhaust emission problem, because as the engine is started and begins to warm up, the cylinder liner, which is directly cooled by a surrounding coolant, expands more slowly than the piston, allowing exhaust gases to escape.

Improved: Although the space that gasoline engine designers leave between the piston and its cylinder allows the piston to expand more rapidly at startup, the space also contributes to the exhaust emission problem. The cylinder liner, which is directly cooled by a surrounding coolant, expands more slowly than the piston, allowing exhaust gases to escape.

Making the Agent and Action the Subject and Verb   Weak prose is often the result of indirect wording, especially through passive verbs. By making your prose more direct, you can often simplify it.

Weak: The cost of the filtration program was found by the design team to be justified, if it resulted in a greater efficiency in the performance of the instrument.

Improved: The design team found that the cost of the filtration program would be justified if the instrument performed more efficiently.

Avoiding Excessive Nominalizing   "Nominalizing" means forming nouns from verbs. When you make the noun "acceleration" from the verb "accelerate," you are nominalizing. Technical prose uses a lot of these nouns, but the result can be hard to understand. Look for ways to make a statement more directly.

Weak: Regeneration of the resin bed is achieved by a calcium chloride solution.

Improved: A calcium chloride solution regenerates the resin bed.

Combining Sentences that Connect or Repeat   In your first drafts, you will often repeat ideas needlessly or fail to link sentences that are logically related. The writing in first drafts is often loose. Combining sentences tightens the prose.

Weak: The main operating cost of the hydro unit is catalyst replacement. Catalyst life has a major impact on the overall economics of the hydro unit.

Improved: Catalyst cost and life largely determine the economics of the hydro unit, because its main operating cost is catalyst replacement.

Substituting Phrases or Clauses for Stacked Modifiers   Modifiers are important when you need to make your writing specific, but they can be overused. Use phrases or clauses to avoid piling modifiers on top of one another.

Weak: This underground plant effluent soil contamination did not threaten employees' health.

Improved: This underground soil contamination by the plant effluent did not threaten employees' health.

Crossing Out Unnecessary Words and Phrases   Most writers draft wordy, convoluted prose, which needs to be trimmed. Cutting the fat from your sentences allows you to communicate economically.

Weak: The cooling of the thermal unit is accomplished by the use of electric fans that are run every other hour during the day.

Improved: The thermal unit is cooled with electric fans every other hour during the day.

Making Vague Words and Phrases More Specific   Drafts are often vague, sometimes because you haven't had time to work out the details or find all the information. Be sure to match the amount of detail to the needs of your audience.

Weak: During the test, the system will be exposed to high temperatures.

Improved: During the test, the system will be exposed to temperatures of 400° to 450°C.

From General to Specific

When you deal first with organization and then with style, you're moving from general to specific parts of your document. Looking first at the big picture means that you will clarify your aims, which in turn will determine many smaller revisions. Your next steps are to analyze your sentences for structure and word choice.

Before you're finished, however, you need to think about illustrations. Graphics, usually labeled as figures or tables, can be as important as prose. They both enhance and summarize information. As you draft any document, you need to think about the visual part of your presentation.

5

Developing Graphics

You've spent months drafting and revising a report. You've carefully considered your audience and defined your aims. You and your colleagues have read and reread each other's work. Your attention to organization and style has made the final version concise and readable. Looking over the finished report, however, you feel something is missing. Paragraph after paragraph passes without a visual break, without graphic support of your argument. You're still not finished. You need to illustrate your prose better.

Of course, you're unlikely ever to be in such a fix. Engineering and scientific writing so often call for visual elements that most writers work on the graphics while they draft the text. Effective graphics condense large amounts of information. They focus attention and reduce data. They are also crucial in analysis and argument. Visual representation has been essential to science—from Euclid's geometric forms to Mendeleev's periodic table to Watson and Crick's double helix. An effective graphic promotes thinking and discussion.

Graphics as Analysis and Illustration

A table, graph, or drawing often begins as an analytical tool to help a writer think about a subject and evolves into an expository strategy to help convince a reader of an argument. The graph in Figure 5.1, for example, shows a mathematical relationship experimentally discovered between the vibration amplitude of welded parts and the strength of the

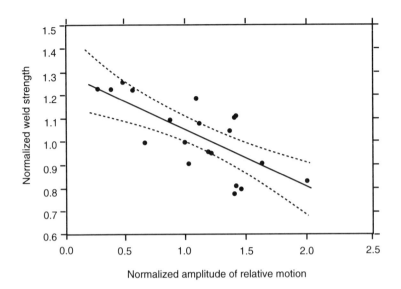

Normalized amplitude of relative motion

Figure 5.1
This analytical graph correlates two variables: welding motion and welding strength. Each variable is normalized at 1 and thus has no terms of expression. A linear fit and 95% confidence limits show the trend and qualify the data. (Source: Jagota and Dawson 1987. Courtesy of ASME.)

weld. The scatter chart, sometimes called a scattergram, correlates two sets of data and reveals a trend.

Graphics are also illustrative. Figure 5.2, for example, shows a schematic and a graph that could dramatically reduce the prose description. The result is to focus attention on essential elements. As analysis and illustration, graphics serve to

• *Study numerical data and physical design.*   Graphic analysis can help you define and think about your subject.
• *Condense information.*   Visuals can summarize information. Use them to concentrate material and to provide overviews of large amounts of data.
• *Improve audience appeal.*   Tables, graphs, and diagrams can clarify information for your main audience and attract audiences that would not normally read your work.
• *Focus the argument.*   Graphics attract attention. Use them to control your discourse. Feature subject matter that's pivotal to your aim.

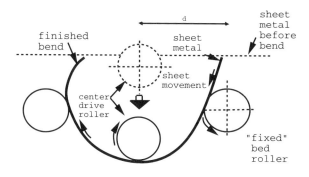

(a)

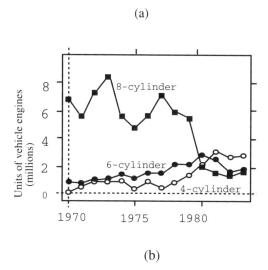

(b)

Figure 5.2
Illustrations focus material and sort information. (a) Schematic illustration of a
sheet metal bending machine. (Source: Hardt et al. 1982. Courtesy of ASME.)
(b) Illustration of trends in auto engine manufacture over a 13-year period.

• *Support the discussion.* Visual evidence often comes closest to demonstrating the phenomenon in question.

Graphics, however, can incorporate the right information without making the right point. Figure 5.3a, for example, attempts to make a specific point about the likely stability of aluminum use by comparing trends for specific materials used in military and civil aircraft. Yet the four pie charts do not facilitate a reader's recognizing either the general trends or the specific instances of aluminum use. The eye has trouble distinguishing the comparative sizes of pie slices in two different pies (not to mention four), and the charts are therefore ineffective for comparing two sets of data. A double-column chart (Figure 5.3b) is a better means for comparing trends, although it places less emphasis on the breakdown for each year.

Developing Graphics by Exploring Data

Developing an effective graphic often means transforming raw data into patterns that advance the discussion. For example, a team is using instruments to monitor changing concentrations of particulate matter in the Beaufort Sea, off the northern coast of Alaska. Drafting their report, team members first tabulate their data according to specific dates and depths, as shown in the table in Figure 5.4. Such tabulation is the simplest way to organize information. We can see the entire 3-month data collection summarized and sorted. Tables preserve data in accurate numerical form. Yet they also sacrifice point of view. This table does not show trends or reveal specific patterns. Hence it may do little to advance discussion.

Tabulation is the first step in preparing all other statistical graphics: line graphs, bar charts, histograms, scattergrams, and the like. These forms, whether developed with a spreadsheet and computer graphics package or worked out on a pad of grid paper, help you examine your data for important trends. Depending on the point you're making, you could cut a variety of graphs out of the data in Figure 5.4. You might, for example, create any of the graphs in Figure 5.5. A line graph could show a trend; a semilogarithmic chart might emphasize rate of change; or a bar chart could emphasize relationships to baseline.

(a) weak version

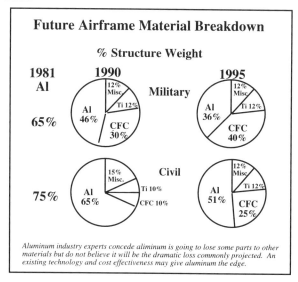

Future Airframe Material Breakdown

% Structure Weight

1981
Al

65%

75%

1990

12% Misc
Ti 12%
Al 46%
CFC 30%

15% Misc.
Al 65%
Ti 10%
CFC 10%

Military

Civil

1995

12% Misc
Ti 12%
Al 36%
CFC 40%

12% Misc
Ti 12%
Al 51%
CFC 25%

Aluminum industry experts concede aliminum is going to lose some parts to other materials but do not believe it will be the dramatic loss commonly projected. An existing technology and cost effectiveness may give aluminum the edge.

(b) improved version

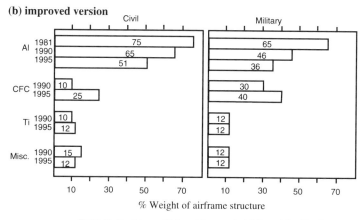

Civil   Military

Al 1981   75   65
    1990   65   46
    1995   51   36

CFC 1990 10   30
    1995 25   40

Ti 1990 10   12
    1995 12   12

Misc. 1990 15   12
      1995 12   12

10   30   50   70   10   30   50   70

% Weight of airframe structure

Figure N. Breakdown of future airframe materials by weight. Aluminum industry experts concede that aluminum is going to lose some parts to other materials, but do not believe it will be the dramatic loss commonly projected. An existing technology and cost effectiveness may give aluminum the edge.

Figure 5.3
Make the graphic support the argument. In (a), the author attempts to use four pie charts to compare component values of four wholes. The results are puzzling. The object is to compare a trend in airframe material use, with special emphasis on alumninum use. In (b), a bar chart with two data fields is used to compare each material, with emphasis on aluminum projections. (Adapted from *Aviation Week & Space Technology*, 1984.)

Heading →

Table 1. Concentrations of total particulate matter, particulate calcium, and particulate aluminum in the upper 100 m of the Beaufort Sea.

Columnhead →
Stubhead →
Column →
Row →
Cell →
Rowstub →

Depth (m)	Apr 10	20	30	May 10	20	30	Jun 9	19	29	Jul 9	19
						Sampling date (1989)					
					Total particulate matter (µg / liter)						
10	49	180	129	86	45	37	38	61	61	44	60
25	83	116	72	78	105	19	30	68	46	44	37
50	132	108	131	77	43	28	32	19	48	34	36
100	24	20	52	52	28	18	21	25	32	24	26
					Particulate calcium (µg / liter)						
10	2.3	11.2	5.4	5.4	0.3	0.3	2.2	2.6	5.4	2.4	3.1
25	3.1	9.1	3.3	3.3	2.4	0.2	1.5	0.8	4.4	2.5	2.5
50	10.5	3.3	3.1	3.1	0.8	0.2	2.1	1.3	4.3	2.6	2.6
100	2.5	16.8	1.5	1.5	0.5	0.1	3.3	3.7	3.1	1.2	3.1
					Particulate aluminum (µg / liter)						
10	0.16	0.34	0.29	0.99	0.31	0.48	0.14	0.18	0.12	0.10	0.14
25	0.12	0.27	0.21	0.88	0.50	0.19	0.13	0.44	0.10	0.13	0.10
50	0.19	0.82	0.17	0.17	0.18	0.10	0.93	0.07	0.05	0.05	0.09
100	0.08	0.21	0.04	0.06	0.09	0.17	0.62	0.12	0.60	0.92	0.08

Figure 5.4
Tables are the simplest visual format and preserve the original data. Each cell represents a full sentence. Tables do not, however, convey visual patterns and may hide significant events or trends.

Determining the Type of Graphic

Graphic design should be unambiguous. Note in Figure 5.6a that a bar chart shows the trends in cost of installation and output costs of photovoltaic and grid electricity in the United States and less developed countries (LDCs). This decreasing cost trend is striking, but the bars are incoherent because the lower scale for "Installed Peak Capacity" has no systematic relationship with the upper scale for "Output." The two sets of values can't be represented by the same bars because each bar is plotted on two scales, which are inconsistent with each other. The author is forced to break the scale several times.

By comparison, Figure 5.6b sorts the information on two scales. Still, the graph is now crowded with notes and the bar of the largest value must be broken to fit on the data field. This weakness is eliminated in the two-way dot chart plotted on a logarithmic scale in Figure 5.7. The information is now presented without broken scales and with additional information. A logarithmic scale, however, is often too abstract for a general or managerial audience.

Data graphics offer an immense variety of options for analysis and presentation. The choices range from simple line graphs that track a set of values for a given item to relatively complex series that plot correlations of two interrelated variables.

Items with Different Values   One of the most common graphic displays is an item with different values. Examples might include the cost over the base price of a specific service, or the average protein content of samples of corn taken at six different latitudes. These differing values could be plotted as dependent variables on the vertical scale of a line graph or bar chart or on the horizontal scale of a dot chart or pictograph (Figure 5.8).

The bar and dot charts, effective for showing discrete values, are the most unambiguous form for this kind of data display. The dot chart is more abstract but avoids masses of ink over specific numerical values. Pictographs often effectively convey information to broad audiences. A line graph, which suggests a continuous series of values, would not be as appropriate. Use line graphs if you wish to emphasize a trend.

Time Series   A time series, probably the most widely used of all data graphs, plots a changing value for one or more items in relation to some unit of time. Hence you might have growth in area of a bacterial culture per 1-hour period or sales volume in millions of dollars for an item per month for a 1-year period. This serial information may be expressed in a line graph emphasizing a continuous trend or in a bar or dot chart emphasizing discrete measurements of a continuous process (see Figure 5.5).

Percentages   Line graphs, bar charts, dot charts, and pie charts can all express percentages (Figure 5.9). Percentages may be parts of a whole, or they may be percentiles, which express more subtle distinctions. For

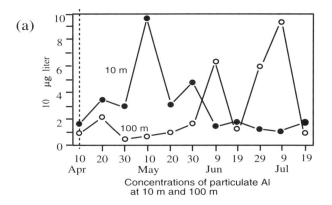

(a)

Concentrations of particulate Al
at 10 m and 100 m

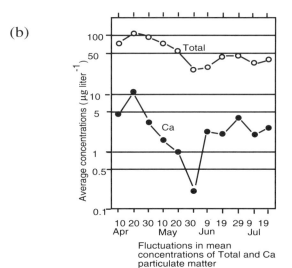

(b)

Fluctuations in mean
concentrations of Total and Ca
particulate matter

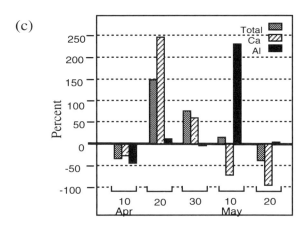

(c)

Percent deviation from mean concentrations of
Total, Ca, and Al particulate matter at 10 m .

example, you could represent the percentage of ozone in air samples at 100-meter increments between 5000 and 10,000 meters, or you could graph the percentage of a specific population or percentile of students scoring A, B, C, D, and E points on a given exam.

If the percentage readings are taken at very frequent intervals, then a line graph is your likely choice of display. For larger intervals, the bar chart is more appropriate. For a simple breakdown of a given sample, the pie chart is the most effective.

Comparison   In plotting comparisons (Figure 5.10), the challenge is to design a graph that displays the trends simply and clearly. Comparative charts can be complex, as the dot chart in Figure 5.10 shows. Pie charts and surface charts are not as effective for comparing values as the line graph, bar chart, or dot chart.

Correlation   Correlations demonstrate or suggest the mutual influence (covariance) of two variables (Figure 5.11a). The variables *A* and *B* are normally two different items or two aspects of a single item. The correlation in Figure 5.1, for example, relates the strength of a weld to the motion of the welded parts during welding. Electrical output is a function of material composition, and bacterial growth is a function of ambient temperature. Correlation represents these linkages.

Each variable depends on the other, or both variables depend on yet another phenomenon. The covariance may be positive or negative, where an increase in *A* shows a proportional increase or decrease in *B*. The independent variable is plotted on the horizontal scale for a line graph or scatter chart. The line graph and scatter chart are most effective for

Figure 5.5
Graphics can manipulate data. Strikingly different graphics may be prepared from the tabular data in Figure 5.4, depending on the writer's purpose and audience. The line chart (a), focuses on the comparative flux of one element at different ocean depths. The semilogarithmic chart (b) plots the mean values of two data series of very different magnitudes. The deviation bar chart (c) illustrates the degree of flux between April 10 and May 20 for all the elements of the original table in Figure 5.4.

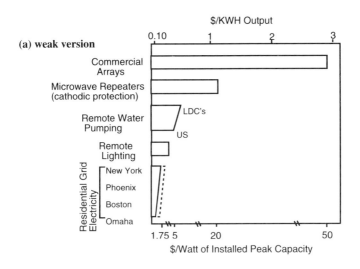

(a) weak version

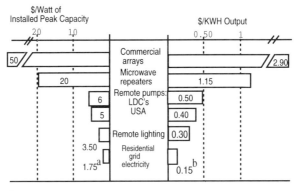

(b) improved version

[a] Average of New York (1.95), Phoenix (1.75), Boston (1.65), and Omaha (1.30).
[b] Average peak load price for New York (0.19), Phoenix (0.17), Boston (0.12), and Omaha (0.10)

Figure 5.6
Clear graphic conception: eliminate ambiguity in scales, data, and labels. The two versions of the bar graph show comparative installation and output costs for photovoltaic and grid electricity. In (a), the author attempts, with confusing results, to plot data on a double scale. In (b), the scales are paired as a way of sorting the information and maintaining clarity.

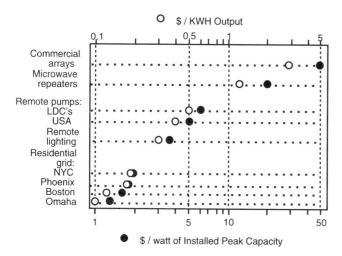

Figure 5.7
A logarithmic scale further resolves the data given in Figure 5.6. The logarithmic scale does not need to be broken to fit into the graph field. You can see the rate of change dramatically here.

displaying correlation because each data point is an expression of the two variables. Softer correlations may be suggested in bar charts and dot charts.

Be careful not to confuse correlation with comparison. Correlated variables are supposed to be interdependent, although proof of their interdependence is often the key issue in the graphic display.

Ratios and Rates of Change   The semilogarithmic scale (Figure 5.11b) is an effective way of plotting data for an item whose values vary greatly. For example, if the series of data points jumps from the 10^1 range to the 10^3 range, the trend cannot be seen clearly on a standard scale because a scale designed for the 10^3 range would not show variations at the 10^1 magnitude. With the semilogarithmic scale, which shows rate of change, different magnitudes may be plotted with great detail and clarity. An increasing slope shows increasing rate of change in relation to prior values.

Line
Graph

Bar/column
Chart

Dot Chart

Pictograph

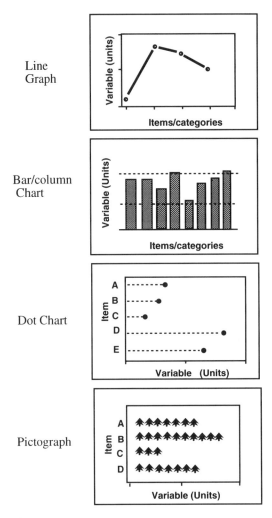

Figure 5.8
Items with different values.

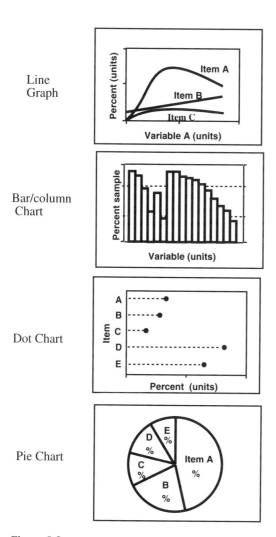

Figure 5.9
Percentages expressed graphically.

Line graph

Bar/column chart

Dot chart

Multiple data fields

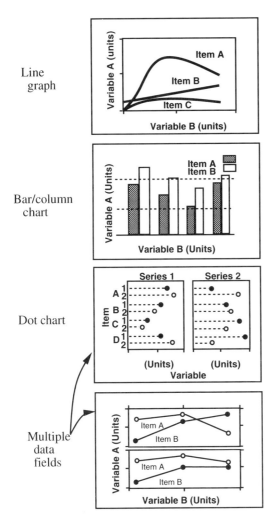

Figure 5.10
Comparisons expressed graphically.

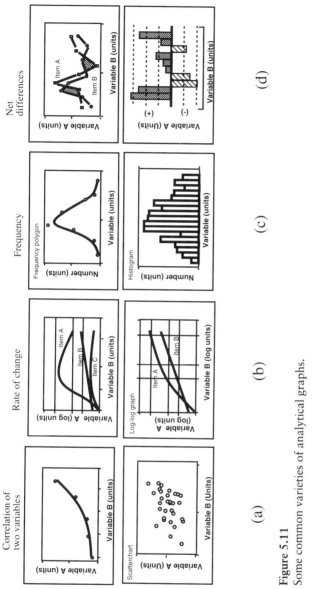

Figure 5.11
Some common varieties of analytical graphs.

Frequency Distributions   Distribution graphics illustrate the spread of a population, as in the classic bell curve distribution demonstrating probability (Figure 5.11c). In constructing frequency graphs, you select an incremental unit (the independent variable) that will best reveal the shape of the sample population and then plot the number or percentage of items in the total population that belong in each increment.

Some variables, such as electric current or heat, are continuous, whereas others, such as population, are discontinuous. Continuous variables are best represented with a line graph, a frequency polygon (with data points connected by lines), or a smooth curve. Discontinuous variables are best presented in histograms or step charts.

Note that in your frequency graphs and charts, the plotting of data will depend on data collection. Data may be plotted for specific values (e.g., $0°$ or $5°C$, Day 5 or 10), or data may be plotted for intervals (e.g., 1–4 or 5–9°C, Days 5–9 or 10–14). For specific times or units of value, plot the data point directly above the value on the horizontal scale. For intervals, plot the data point above the midpoint of the interval on the horizontal scale. For a step chart or histogram, the interval (i.e., 5–9 days) is the width of the column.

Net Differences   Net difference charts use both positive and negative values to show fluctuation and resulting differences (Figure 5.11d). The deviation bar chart in Figure 5.5c illustrates the percent deviation from the mean concentrations of the measured particulate matter in samples from the Beaufort Sea.

Designing Graphics

Designing and preparing graphics takes time, even with the powerful assistance of computer graphics and spreadsheets. Crude draft graphics should suffice until you're ready to commit the time needed to complete the formal version. Preparing a graph or chart involves a series of steps:

1. Select a design that demonstrates the point you want to make.
2. Prepare and label the axes. Place the independent and dependent variables as follows:

	Variable on the horizontal axis	Variable on the vertical axis
Tables	Independent	Dependent
Line graphs	Independent	Dependent
Bar charts (hor.)	Independent	Dependent
Bar charts (vert.)	Dependent	Independent
Dot charts	Dependent	Independent

3. Design scales that best reveal the data.
4. Plot and label the data.
5. Prepare an informative legend.

Adapting Graphics to Your Readers

Readers vary greatly in their familiarity with graphics. A specialist in a scientific or technical field normally expects a range of interpretive analytical graphics—scattergrams, histograms, logarithmic charts, detailed design drawings, and the like. The schematic shown in Figure 5.12 is a typical no-frills illustration of an apparatus addressed to a technical audience. By contrast, administrators often need much-simplified graphics that address their managerial concerns for understanding a situation and allocating resources. Simple line graphs, bar charts, pie charts, diagrams, and tables are appropriate for such decision making.

Making Graphics Consistent with Your Data

Visuals may distort information if they don't accurately reflect your data. You can see this problem in the line chart of Figure 5.13a. Trends for solid fuel consumption in the United States, Western Europe, and the former Soviet Union are shown over the period 1965–1976, but with only three data points for each trend, so that the line graph is deceptive. It suggests data for the 9-year gaps.

This weakness is eliminated in the grouped column and dot charts (Figure 5.13b,c). Visual representations are now given only to the years for which data exist; trends are only suggested. The bar chart is a traditional form appropriate for managerial and general audiences. The dot chart is less cumbersome.

Plotting lines for trends in discontinuous data is not always wrong. Be sure, however, that the variation in the gaps between data points cannot

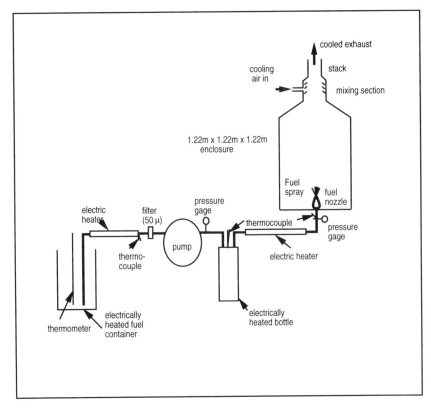

Figure 5.12
A typical schematic demonstrating a process mechanism with flow moving from left to right. The apparatus is an experimental setup for atomizing No. 6 oil. (Source: Kwack et al. 1992. Courtesy of ASME.)

easily be greater than the variation between any two data points in the line chart.

Double-Checking the Accuracy of Curves and Scales
The accuracy of data is always an issue. At the writing stage, authors face two special concerns: technical accuracy and presentational accuracy. Technical accuracy concerns the data and related analytical methodologies. Presentational accuracy concerns the visual representation of data with a minimum of distortion. Some of these issues are illustrated in Figure 5.14.

(a)

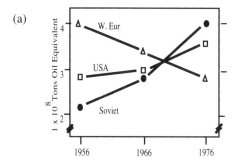

(b)

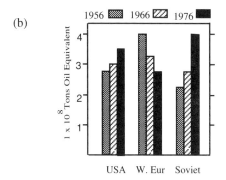

(c)

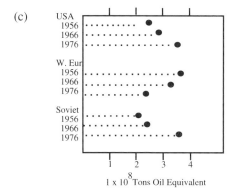

Figure 5.13
Your graphic should be consistent with your data. Here the same information for solid fuel consumption is placed in three different graphic formats. Although the line graph (a), grouped bar chart (b), and grouped dot chart (c) are equivalent expressions of the data, the line graph is less appropriate because it plots data for all years, including the 9-year gaps.

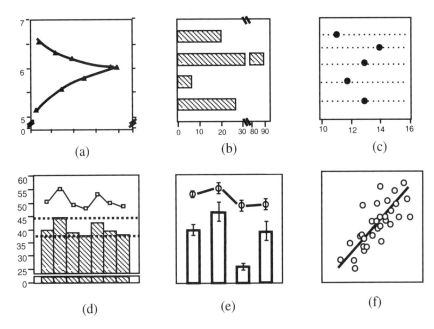

Figure 5.14
The scale and design of a graph can affect accuracy. The horizontal and vertical scales for line graphs should begin at zero. A suppressed zero (a) or scale break (b) can be used to eliminate a large blank space. A continuous horizontal scale line across a dot chart (c) eliminates the need for a zero baseline. Scales should clearly reveal variation without exaggeration (d). Use range bars (e) to show standard deviation, standard error, or confidence intervals—be sure to indentify which! Theoretical curves and lines with mathematical formulas (f) should be plotted along with emperical data for comparison.

Integrating Graphics with Text

While a graphic needs a clear visual context established by caption and labels, it also needs a clear textual context established by discussion. Titling and labeling take some effort.

Descriptive labeling should echo textual references that help the reader understand the graphic's significance. Tell the reader how the visual advances, supports, clarifies, or summarizes your discussion. The following techniques help integrate graphics with text:

• Tell the reader why the graphic is important.
• Tell the reader what the graphic means.

• Explain how the graphic supports the main discussion.
• Add parenthetical information (conditions, limitations) not identified in the graphic.

Preparing Final Graphics

Because fine details can make the difference between a graphic with an audience of one and a graphic with an audience of many, you want to be doubly vigilant as you prepare final versions of your visuals. Cryptic and incomplete labeling are among the most common sources of ineffective graphics. Emphasize data, not graphic apparatus. Don't use the fancy visual background patterns available on many computer graphics routines, for the apparatus can easily overwhelm your information. Be sure, however, that your graphic apparatus is thorough and visible.

For example, a crude graph, such as the line graph of Figure 5.15, will have many minor problems that need attention in the final stage of preparing the manuscript. The example is a rough semilogarithmic line graph prepared from the table of values in an experimental series. The graph is numerically accurate but needs cleaning up and resizing for final production. If the graph were reduced to 60% of the original, as many graphics routinely are in final production, the information would disappear. The improved version is shown in Figure 5.16.

Consider the following as you revise (see Figure 5.15):

1. *Legend placement.*   Legends should be part of the text, not part of the figure. If you are submitting your figure for outside publication, include the figure number and legend in brackets where you think it should appear in the text. Tape the number on the back of the graphic. For final publication, place figure legends at the bottom and table titles at the top of the page.

2. *Symbols.*   Check to make sure that you are using the standard units and symbols of your discipline.

3. *Legend wording.*   Wording should be unambiguous. Place modifiers next to words they modify.

4. *Data points.*   Be sure that data points are large enough not to be obscured when reduced. Avoid placing them on the box, where they can be missed. Use squares, circles, and triangles to distinguish among curves. Emphasize actual data points and make lines between them less prominent. When the curve itself is a calculation, make it the most prominent line on the graph.

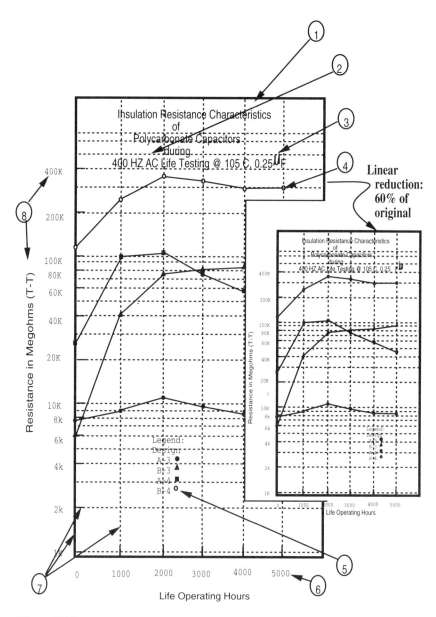

Figure 5.15
A draft graph. Although the information is accurately plotted, the graph is poorly sized for reduction, the labeling is unclear, the data region is filled with clutter, and the data points are too small for later reduction for publication.

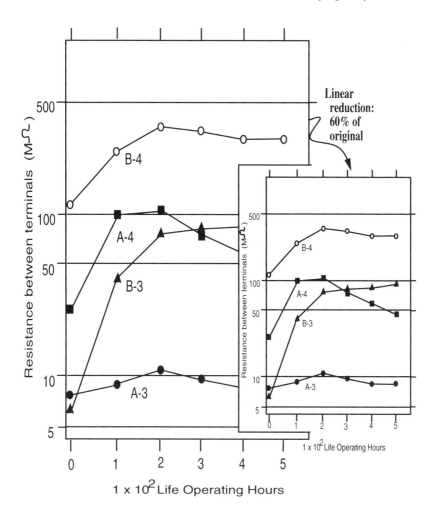

Figure N. Insulation resistance characteristics of 0.25 μ F polycarbonate

capacitors (400-Hz ac life testing at 105 C°).

Figure 5.16
This revised version of Figure 5.15 has improved the clarity of the information. The original legend and labels have been reworded. Note also the emphasis on data points and curves.

5. *Labels.*   Whenever possible, just label parts of the figure (e.g., curves). Avoid extra comments that clutter your data field.

6. *Scale numbering.*   Keep the scale units simple, with unit multiples (e.g., 1000s), just below or beside the scale.

7. *Boxes, tick marks, and grids.*   Locating aids help orient the reader, but use them with restraint. Keep them thin and unobtrusive. If you use a box around your graphic, place tick marks outside the box, where they do not compete with the data.

8. *Nonstandard abbreviations in labels and legends.*   Avoid using cryptic abbreviations. When abbreviating, include enough letters to assure clarity. For example, a "T" for "terminal" should be spelled out in the legend so as not to be read as "temperature."

The Many Uses of Graphics

Graphics are key components of most documents—proposals, reports, articles, and entire books—where they arrange and summarize data. Plan a graphics strategy at the early outline stage of your document. Visuals are also essential to many oral presentations, both formal and informal. You may find that an audience will focus attention best on a chart or graph or that a diagram will readily explain what words can obscure.

6
Conducting Meetings

As a project manager in a large R&D company, you are charged with producing a document describing a proposed new product. The final document will span several volumes. It will include research findings, backup documentation, manufacturing plans, and quality assurance data. People from several departments, together with a few subcontractors, will need to write and review drafts. How do you begin?

Chances are you'll call a meeting. With such a large team and so much to do, you'll want representatives from every department involved. Managers will need to know what data their departments are to provide. Writers will need to have their tasks defined. They'll need deadlines. The team will need to determine a review process. While some of these tasks can be accomplished in a meeting, others cannot. Knowing how to use meetings to manage a complicated process may be the key to producing a complete document on time.

Meetings as Collaboration

Because writing is part of a larger continuum of communication, writing for science and engineering is always collaborative. Meetings are important vehicles for collaboration both at the start and throughout the duration of a project. Whether you are a principal investigator coordinating research or a design engineer developing a new component, you need to enlist the help of others to meet your own objectives. Help can, in turn, save you from costly delays and misunderstandings.

At the start of a project, when many questions are still open, meetings are a forum for defining and reviewing problems, developing strategies, exploring methods, and critiquing results and documents. As a project progresses, meetings are a unique means of communication. Effective meetings encourage the sharing of expertise and responsibility among colleagues contributing to

- Problem identification
- Consensus building
- Group decision making
- Information exchange
- Document review
- New ideas

Meetings can also be notorious time wasters. Group behavior and interactive speech are often amorphous, and meetings do not easily focus attention, assume direction, or deliver concrete results. Everyone's agenda is normally filled with tasks demanding immediate attention. Most people see meetings as unwanted diversions. Team leaders—or anyone else chairing a meeting—need to make the effort worthwhile.

Meetings serve both motivational and structural purposes. To motivate others, the chair needs to establish purpose and context to avoid rambling discussions, personality problems, and poor attendance. To structure meetings, the chair needs to work from a clear agenda, establish effective time limits, and develop means for follow-through.

Managing Collaborative Writing

When collaborative writing fails, it is most often the result of misunderstandings or disagreements over problem definition, research procedure, writing responsibilities, scheduling, and manuscript reviews. Most of these problems can be minimized by strategic use of meetings. Meetings provide a vehicle for accountability, which prevents writers from falling too far behind. Meetings also facilitate the circulating of drafts among colleagues and across hierarchies before submission to clients, agencies, or journals.

Collaborative writing is often organized around meetings when a group has been assigned to produce recommendations. Even informal collaboration requires two or more people working together to produce a document. Meetings are inevitably part of that process. Still, group writing is difficult, and group process is often confusing.

Peer Collaboration   Meetings for collaborative writing projects are most productive when the writer has circulated documents in advance and plans a way to record the group's findings. This review process may be carried out electronically by means of e-mail routing or posting and annotation. Collaboration may also take place individually. One-on-one, the main writer has greater control over the way in which responsibilities are assigned, scheduled, and met; meeting time is minimized while each individual can still contribute.

Electronic routing of a document is the handiest—and probably the most common—form of collaboration. Document routing requires little planning and very little meeting time. One writer drafts the document and sends it to coworkers via e-mail. Coworkers provide critiques either by annotating the document or by directly making revisions in high-lighted text. Most word-processing software supports these activities.

If the document is a large-group effort, such as a proposal or a major report, then a committee, with regular meetings, is likely to be necessary. Following are some guidelines for using meetings in organizing and carrying out writing projects:

• *Organization.*   Prepare and use outlines as control documents. An outline helps the main writer get agreement on scope and approach. Without such controls, groups are difficult to keep organized, especially when each member is producing part of a larger document. An outline facilitates the assigning of different responsibilities to different people.

• *Style sheets.*   Establish basic formatting and documentation conventions, either by adopting a standard such as the *IEEE Style Sheet* or a word-processing document template or by drafting your own format for subject headings, numbering systems, word usage, figures, and so on. The simplest approach is to designate a published document or document template as a standard and use it as a guide for stylistic consistency.

• *Schedules.*   Schedule meetings as a way of establishing milestones for a draft. Agree on deadlines so that the project is not held up by straggling contributors and reviewers. Otherwise, you have no way of knowing whether your collaborators are producing their writing. Frequently, you will find, they are not.

• *Review.*   Establish a review mechanism, whether in electronic form or in hard copy, so that everyone is aware of the process. Make the principal author or editor responsible for coordinating the review. Meetings may be effective for gathering a range of views, but some writers prefer individual collaboration. Not everyone is comfortable receiving critical commentary in a group setting.

Supervisory Collaboration   Group writing can take a hierarchical form, especially in organizations. In these settings, supervisors review the writer's work both for its technical accuracy and for its institutional implications. Supervisors typically review assertions and recommendations for the way they reflect the policies of the work group and the larger organization.

Seniority or authority characterizes supervisory collaboration, as the supervisor can require the writer to make certain changes. For example, a report assessing how effectively a contractor is meeting the terms of an agreement might contain much criticism. The writer may feel that the critique is justified, whereas the supervisor may feel that the criticism is harsh and antagonistic. Differences in perception are common to all collaborative writing, but a hierarchical relationship can make collaboration potentially abrasive.

Once again, meetings are an effective means of discussing differences of opinion and reaching a preliminary understanding. In supervisory reviews, all parties benefit from early agreement, before the writing has proceeded so far that the writer has trouble carrying out the revisions. As Figure 6.1 suggests, Collaboration is much more effective when the parties achieve early agreement because, as writing progresses, the writer invests more time and identifies more intensely with the work. Personal ego is increasingly at stake, and required revisions become harder to swallow. The object is to avoid situations in which a writer thinks a document is finished but must then extensively revise it.

Structuring Meetings

Meetings require care in planning and conducting. Essential to success is an effective structure that will provide an amorphous oral process with clear organization, consistent focus, and the means for follow-through.

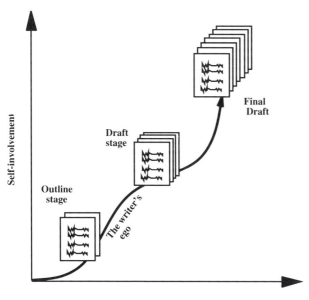

Time, concept development

Figure 6.1
The writer's growing resistance to change. Both writers and supervisors should seek early consultations, preferably at the outline and draft stages. As writers progress with a document, they invariably become ego-invested in their writing. This increasing involvement suggests that collaborative revisions are much easier in the early stages of writing.

Plan the Meeting   Neglect in planning meetings is probably the most common reason they fail. To plan a meeting, you have to get everyone to agree on a meeting time and place at least 5 to 10 working days before the meeting. You also need to inform participants about meeting length, place, and subject.

If the meeting is informal, you can arrange these matters over the telephone, although reiterating the arrangement in a brief follow-up note is still a good idea. If your meeting involves a group or a formal committee, you need a written announcement stating the place and time of the meeting, its main purpose, and the items on the agenda (Figure 6.2).

An agenda should progress from (1) routine, context-setting items to (2) general information discussions that do not require decision making to (3) the main decision-making discussion to (4) recapitulation and

MEMORANDUM

To:
From:
Subject:
Date:

The meeting called by J. Aggarwal to discuss general interest in establishing a new Energy Laboratory spectroscopic facility will take place as follows:

Thursday afternoon, December 4, from 3-4:30 pm
in Room 3-337 of the Ames Research Center.

The main purpose of the meeting is informational. We are interested in identifying level of interest and potential sources of technical and financial support. We will determine at this meeting whether to pursue this project at the facility-wide or local levels.

Agenda:

o Introduction -- J. Aggarwal

o Other national laboratories and their approaches--S. Hunt

o Some staffing and computing requirements--L. Dickson

o Possible local sources of technical and financial support--J. Aggarwal

o How to proceed in the next phase--Discussion

    o Should we handle this as a facility-wide or as a local
      laboratory need?

    o Should this need be addressed next year, or should we wait to see
      what happens in our sponsored funding patterns?

o Other items for future agenda

The discussions will be preliminary and the only decisions we will make at this meeting will be to answer the two questions concerning proceeding.

Please be on time.

Figure 6.2
A sample agenda. Note the combination of assigned presentations and decision-making discussions. Such detailed agendas improve participation by providing participants with a chance to review items before the meeting.

Minutes into the meeting

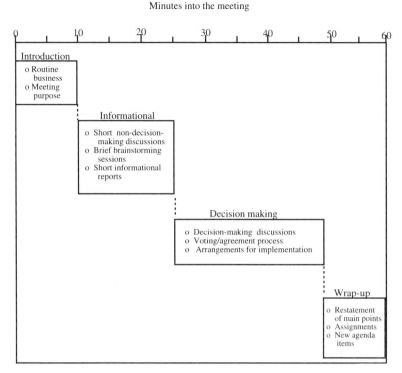

Figure 6.3
Timing segments for a 1-hour meeting. Start slowly and build to the core discussion. Be careful not to overload the agenda. Only five 10-minute items can be discussed in an hour.

assignments (Figure 6.3). Circulate the agenda at least 5 to 10 working days before the meeting. This way, you keep everyone informed, let people think about issues before the meeting, and let participants know that your plan is serious.

Work from a Written Agenda   Conduct meetings from written agendas. An agenda should be of reasonable length, not so long that your meeting ends halfway through the items. The agenda, normally prepared by the person who calls the meeting, provides guidelines for conducting the meeting and keeping it on course. Generally, the chair refers to the agenda throughout the meeting, spends a given amount of time on each item, and brings the discussion to a close. Meeting management is crucial to success.

Maintain the Meeting's Focus   Routinely establish and reestablish your context. Restate the topic and remind participants of the issue under discussion. People will readily stray into other topics, some important and some irrelevant. The chair—or even interested participants—should bring a straying discussion back to the agenda. If participants want to move into new productive territory, reserve time for the topic at the next meeting or allow discussion at the end of the meeting. You need to do this diplomatically by appealing to time and agenda constraints.

Maintain Momentum   Set goals and strive to reach them. Meetings need to progress, so that the agenda is covered adequately. Time the meeting, allotting each item so many minutes, and then move on. Digressions quickly wreck meetings. A 1-hour meeting can cover four to five topics if you are planning to allot 10 minutes to each one. Expect to take 5 to 10 minutes to get the meeting underway and to frame the discussion. Effective meeting dynamics require that the chair view each discussion as part of a whole and move the group on at a steady pace.

Establish a Record   Loss of content is a common problem. Memories of discussions soon fade, and entire meetings can be consigned to oblivion because no one has jotted down a record of the main discussion points and the decisions reached. Taking notes during or immediately after the meeting establishes a record of ideas, names, and agreements. If the meeting is informal, each member might record notes in a personal notebook. If the meeting is formal, the chair needs to designate an administrator who will take notes and prepare minutes for later reference.

Minutes should summarize the main points discussed and the decisions made. Only occasionally is precise wording necessary, and then only if the wording is important. Minutes are not normally verbatim records; tape recorders do that better. The minutes for an hour-long meeting, for example, will not normally run more than one to one and a half pages. Minutes need to be submitted for the approval of the participants.

Keep Participation High   Encourage everyone to engage in the meeting. Don't allow ten people to attend a meeting where three participates while seven others sit and say little or nothing. People unwittingly collaborate in these silences. While some are content to let others talk, others are happy to do the talking. The result may be low participation, narrow use of available expertise, and loss of consensus.

Promote participation by studying and assessing personalities and work styles. Call on silent participants and neutralize excessive talkers diplomatically. One effective technique is the roundtable query, in which every member of the group is asked to respond to a question. Also effective is calling on those with specific expertise to prepare brief presentations.

Monitor and Promote Follow-Through   Perhaps the most difficult meeting task is to translate decisions and commitments into concrete actions. Follow-through requires (1) that a written record be established and (2) that people be encouraged to take action on agreed-upon decisions. Minutes are essential for preserving group memory. For more formal meetings, minutes should summarize the main points of discussion and record actual decisions.

Every meeting decision should be followed by some discussion of the means and schedule for the project. The chair or supervisor should contact participants to remind them of deadlines. In formal committees, members should prepare presentations in advance, so that they may be circulated with the agenda for the next meeting. A tightly managed follow-through process is often what distinguishes an effective committee from a loose talking group.

Committees and Collaboration

Committees have become essential to collaboration within and among organizations. Committees work to transmit goals, share expertise, develop consensus, and delegate decision making. Committees must manage the same issues that can derail other collaborative groups, but most committees also engage in a series of meetings.

Managers must carefully define the mission of a committee and the authority that the committee's recommendations will carry. Avoid forming bogus committees around nebulous problems and tasks. Committees will normally balk at rubber-stamping an administrator's decisions and may even produce results different from what the administrator had expected. The ignoring of a committee's recommendations may be a source of infighting and cynicism.

The committee chair needs the authority and standing to require members to work effectively on the committee, and both the manager and the chair need to consider the kinds of assistance that will be useful. The chair is ultimately responsible for setting goals, establishing an agenda, maintaining momentum, keeping everyone involved, monitoring follow-through, and communicating results.

Either an assistant chair or a staff secretary should take minutes of meetings, which become a history of the process. Minutes are then available for reference when needed. Also important to a committee's functioning are setting up meetings, calling committee members, and disseminating information.

Effective time management is essential. Committee members should have the agenda at least 5 working days before the meeting. Reading the agenda in advance enables members to come prepared with ideas. A standard meeting time is usually best. Meetings should start and end on time. If you're willing to start late, participants will come late. If you're willing to end late, members will resent the use of their time for committee purposes.

Committee members stay engaged if they have tasks to accomplish. Keep people involved by delegating responsibilities and assigning tasks at meetings. These tasks should be noted in the minutes, with any necessary deadlines.

Disband the committee when it is no longer useful. Like any enterprise, a committee can lose its reason to exist. Many committees continue to meet long after they have ceased to be useful. In these cases, time is lost, members become irritable, and members cease to be productive.

Meetings in Context

The effort required to run a meeting or committee is substantial; hence organizations should use meetings judiciously. Calling a meeting that lacks a clear agenda, failing to define tasks, or using meetings to avoid the hard work of drafting copy are common mistakes. When planned and led well, meetings can generate new ideas and promote collaboration. But meetings are not always the best way for people to work together. Collaborative communication can take other forms. Memos and mail—both on paper and online—are sometimes the best means of communicating with colleagues.

7

Memos, Letters, and Electronic Mail

As a newly hired manager for an engineering project team, you spend your first day on the job reading the files. You have little time to interview team members and ask follow-up questions. By the end of your first week, you must report to your manager with a detailed proposal. Your assessment of both your project and your employees will depend on team members' correspondence. Memos and letters will form your first impressions.

Memos and letters are brief and relatively informal documents, unlike proposals or completion reports. Yet many technical professionals spend more time writing (and reading) these familiar forms than they spend on any other communication task. Despite their brevity and relative informality, memos and letters may be archived and reviewed later. They may be read by those not originally addressed. Both forms may become important parts of a project record. They may serve as the basis for important decisions, with effects as significant as those of multivolume proposals or formal completion reports.

The structure of both memos and letters is flexible enough to be useful for a wide variety of purposes, including proposals, requests for information, progress reports, notices of meetings, complaints, inquiries, records of telephone conversations, or calls for meetings. The personalized forms of memos and letters distinguish them from other technical workplace documents. They name the recipient, they name corecipients, and they identify the author.

The Differences Between Memos and Letters

The memo form is used for communicating *within* an organization, never for an outside audience. The letter is used for communicating *outside* an organization. Thus a feasibility report prepared for exclusive use within a company will be accompanied by a *memo* of transmittal, and a report prepared for a client will be covered by a *letter* of transmittal. Social practices will vary, of course. A supervisor wanting to congratulate an engineer for having a paper accepted for publication might send a letter to a home address, rather than a memo through the interoffice mail. And in e-mail communication, no distinction is made between memo and letter or between files that will be transmitted to the next office and files that will be transmitted across the country.

Analyzing Your Audience

In shaping the content of memos and letters, you must address the information needs of the recipient. This task is not very challenging if the point of your correspondence is to offer a raise or an extra month's vacation. But many memos and letters are written to ask someone to do something for you: to hire you, to accept for publication an article you are submitting, or to give you an extension on a deadline that has long passed. You will need to ask yourself how the recipient of your correspondence can best be persuaded to adopt your view of the subject—or at least not be put off by it.

In your search for a persuasive strategy, consider what your reader already knows about the situation you are addressing. Ask yourself how this reader is likely to react to what you are saying. Then remember that the first audience for memos and letters may not be the last. If copies of your document need to be sent to other readers, you should also consider how each one is likely to respond to what you have written.

Memos: Format, Organization, and Style

Memo Headings

The elements in memo headings communicate technical and social information. Though the exact placement of elements in the heading of memos will vary from organization to organization, the content remains constant: memo headings invariably identify date, recipient, author, and subject. Memo headings perform important reference functions. The prominence of the date provides a chronology for the issue under consideration, so anyone can see at a glance where each document fits into the evolving life of a project. The date locates each action and may be important later if, for example, you are involved in legal action. Organizational titles and levels of responsibility may influence the relative weight a reader will give each communication. Although scientists and engineers should be influenced primarily by objective evidence, readers are, nevertheless, often influenced by the professional rankings of authors and audiences (Figure 7.1).

Of all the elements of a memo, the subject line carries most responsibility for flagging readers. Because they function as title and abstract combined, subject lines need both to present a concise statement of the memo's topic and to contain information that will tell a reader whether the memo is immediately important. An additional audience for the subject line is the clerical personnel who file your document. They are likely to make filing decisions based on mechanical searches for keywords. An ambiguous subject line can keep your memo from reaching the right reader at the time you send it and later on as well (Figure 7.2).

Memo Text

Each memo or letter you write should adhere to some broad outlines, but within those outlines you develop strategies for organizing and presenting your content to a specified audience. Though the external forms of memos and letters are rigid, the content is extremely malleable. Once you identify your purpose and audience, you can shape your text more precisely than for other technical documents.

A three-part organizational plan works well for most memos. Open with an overview. Tell readers exactly why you are writing and what they will gain from reading. Use the middle section of the memo to develop your point and provide supporting arguments. Use the final section to summarize your point and, when appropriate, to request or suggest follow-up action.

Memos are utilitarian forms, less formal than letters. Memo readers are never addressed as "Dear"; memo writers never conclude with

Internal Correspondence

General Specifics
Research Corporation
Santa Rosita, California

TO: PETER CAMPBELL
             DIRECTOR, RESEARCH AND DEVELOPMENT

FROM: TERRENCE MARSHALL *TM*
             ASSISTANT DIRECTOR, ELECTRONIC MATERIALS LAB

DATE: MARCH 18, 1998

SUBJECT: FUNDING REQUEST FOR GALLIUM ARSENIDE LINEAR
             INTEGRATED CIRCUITS

I was glad to receive your request for proposals. I strongly urge a committment to research in the field of ultra-high mobility Gallium Arsenic (GaAs) linear integrated circuits (IC).

GaAs is quickly becoming the material of choice for IC manufacturing, since:

- GaAs will support transistor switching rates in the gigaHertz range.

- Further miniaturization of IC arrays may be feasible due to the high carrier concentration of GaAs.

- GaAs has exhibited highly predictable behavior over a wider range of temperatures than conventional IC materials.

We already have many of the necessary facilities for growing and testing GaAs crystals for use in IC arrays. With proper funding, I could immediately embark on an intensive study of GaAs here at the Electronic Materials Lab. The success of this project would greatly benefit GSRC and secure a strong position for us in this rapidly developing industry.

I can be reached at extension 3-9050 if you have any further questions.

TM:an

Figure 7.1
This memo heading contains four requisite elements: name of recipient, name of sender, date, subject. The subject line is focused and specific; the body of the memo, with its bulleted list, is designed for rapid reading.

Generic Subject Lines	Informative Subject Lines
Re: Proposal	Re: Request for Support of Hydroponics Project
Re: Meeting	Re: Meeting 3/6 to Discuss Desalination Cost Overruns
Re: Problem	Re: Shipping Delay Stalling GaAs Linear Development
Re: Recommendation	Re: Recommendation to Develop Melanin-Like Polymer as Passive Solar Energy Coverter

Figure 7.2
Informative subject lines contain concise statements of the memo subject, giving readers a helpful preview of content. By comparison, generic subject lines do very little to address the information needs of potential readers.

"Yours truly." In most organizations, they sign their documents with their initials only. But memos are also personal and cordial: by all means, use "I" and "you." A memo is an intraorganizational document, and pomposity or even much formality is not expected. Aim for a style that is efficient and cordial.

But keep in mind that despite their in-house status, memos may become important parts of historical archives. You may be tempted to include a private communication in technical memos; for example, you may want to use the occasion of reporting progress on a new stack gas emission control to add congratulations on the birth of a baby. Yet the personal rarely seems appropriate months or years later. Remember that your memo may need to be reviewed. Many writers attach removable notes to memos and use those spaces for personal comments that they would *not* want retrieved at a later date.

Extensive postmortem analysis of the memos connected with the accidents at the Three Mile Island nuclear plant and the space shuttle *Challenger* has focused attention on the limitations of written language as a reliable way to avoid crisis. Figure 7.3 is a now-classic example of a poor memo. Because this writer failed to impress the reader with the urgency or even the point of his memo, a relatively minor glitch led to a partial core meltdown and release of radioactive gas into the environment. The entire nuclear industry was dealt a near-fatal blow by public and

---

BABCOCK & WILCOX COMPANY
POWER GENERATION GROUP

To   B.A. Karrasch, Manager, Plant Integration

From   D.F. Hallman, Manager Plant Performance Services

Cust.   Date   August 3, 1978

Subj.   Operator Interruption of High Pressure Infection (HPI)

References 1 and 2 (attached) recommend a change in B&W's philosophy
for HIP system use during low-pressure transients. Basically, they
recommend leaving the HPI pumps on, once HPI has been initiated,
until it can be determined that the hot leg temperature is more than
50 F below T_{sat} for the RCS pressure.

Nuclear Service believes this mode can cause the RCS (including the
pressurizer) to go solid. The pressurizer reliefs will lift, with a
water surge through the discharge piping into the quench tank.

We believe the following incidents should be evaluated:

1.   If the pressurizer goes solid with one or more HPI pumps
        continuing to operate, would there be a pressure spike before
        the reliefs open which could cause damage to the RCS?

2.   What damage would the water surge through the relief valve
        discharge piping and quench tanks cause?

To date, Nuclear Service has not notified our operating plants to
change HPI policy consistent with References 1 and 2 because of our
above-stated questions. Yet, the references suggest the possibility
of uncovering the core if present HPI policy is continued.

We request that Integration resolve the issue of how the HPI system
should be used. We are available to help as needed.

                                          _____
                                          D.F. Hallman

DFH/feh
attachments

Figure 7.3
Memorandum from D. F. Hallman to B. A. Karrasch, August 3, 1978. (Source:
Gorinson et al.)

governmental reactions to an incident that a better-written memo *might* have averted. The moral, however, is not simply that bad news must be expressed in vivid, unmistakable language; not all responsibility can be placed on the writer. At least one of the *Challenger* memos is a model of urgency and clarity. Neither the subject line nor the text should have left any reader in doubt (Figure 7.4).

Letters as Both Personal and Official

A letter is simultaneously highly personal and official. You speak directly to the reader named in the opening salutation with the salutation "Dear," and you close the document with your handwritten signature. At the same time, the letter may bear your company letterhead and highlight your administrative level. For example, a typist's initials at the bottom of the page signal to your reader that you are important enough to have secretarial assistance. And when you include the full title and organizational address of your recipient, you indicate that your letters are both written and received in full recognition of institutional hierarchies.

Letters written on institutional letterhead are official forms, and they relay the weight of your office and affiliation. Because communication on company letterhead carries an implied official endorsement, take care when you use it. You are, in effect, expressing not only your own message but also the views of the institution.

Most organizations have a "house style" for letters, with standards for indentation, spacing, and punctuation. The widely used block style is both attractive and functional (Figure 7.5). Though a subject line is not absolutely required, it provides a preview for the recipient and filing information for a clerk who may need to retrieve the letter at a later date. Some organizations prefer modified block. In this style, paragraphs are indented, and date, closing, and signature are aligned approximately two-thirds across the page.

We think that letters should always be addressed to someone, never "Dear Sir" or "To Whom It May Concern." If you do not know the name, title, and preferred form of address of the person you're writing to, you should not, except in unusual circumstances, be writing a letter. Instead, you should be finding answers to these crucial questions. Even

MORTON THIOKOL, INC.

Wasatch Division

Interoffice Memo

31 July 1985
2870:FY86:073

TO:	R. K. Lund
	Vice President, Engineering
CC:	B. C. Brinton, A. J. McDonald, L. H. Sayer, J. R. Kapp
FROM:	R. M. Boisjoly
	Applied Mechanics - Ext. 3525
SUBJECT:	SRM O-Ring Erosion/Potential Failure Criticality

This letter is written to insure that management is fully aware of the seriousness of the current O-Ring erosion problem in the SRM joints from an engineering standpoint.

The mistakenly accepted position on the joint problem was to fly without fear of failure and to run a series of design evaluations which would ultimately lead to a solution or at least a significant reduction of the erosion problem. This position is now drastically changed as a result of the SRM 16A nozzle joint erosion which eroded a secondary O-Ring with the primary O-Ring never sealing.

If the same scenario should occur in a field joint (and it could), then it is a jump ball as to the success or failure of the joint because the secondary O-Ring cannot respond to the clevis opening rate and may not be capable of pressurization. The result would be a catastrophe of the highest order—loss of human life.

An unofficial team (a memo defining the team and its purpose was never published) with leader was formed on 19 July 1985 and was tasked with solving the problem for both the short and long term. This unofficial team is essentially nonexistent at this time. In my opinion, the team must be officially given the responsibility and the authority to execute the work that needs to be done on a non-interference basis (full time assignment until completed).

It is my honest and very real fear that if we do not take immediate action to dedicate a team to solve the problem, with the field joint having the number one priority, then we stand in jeopardy of losing a flight along with all the launch pad facilities.

R. M. Boisjoly

Concurred by:

J. R. Kapp, Manager
Applied Mechanics

Figure 7.4
This memo failed to serve as warning of impending tragedy, this time on the *Challenger* space shuttle. Unlike the memo displayed in Figure 7.3, this memo gets immediately and vigorously to the point. The language in the subject line is straightforward, as is the first sentence and the heart-wrenching last sentence. There are situations that require more intervention than a well-written document. (Source: US Presidential Commission 1986.)

STEPHEN CHUNG AND ASSOCIATES
PROFESSIONAL CONSULTING ENGINEERS
18886 Hollister Ave.
Santa Rosita, California 93069

March 7, 1994   *4 returns*

Mr. Nicholas Altenbernd, President
Quantum Design
18 Wright St.
Cambridge, MA 02139   *2 returns*

Dear Mr. Altenbernd: *2 returns*

SUBJECT: Topic Agenda for Year-End Progress Meeting   *2 returns*

_____
_____
_____.

_____
_____
_____:

  • _____
  • _____
  • _____

_____
_____
_____.   *2 returns*

Yours sincerely,   *4 returns*

Denise Belanger

Denise Belanger
Director of Research   *2 returns*

db   *2 returns*

Enclosures (2)   *2 returns*

c: David Neuman

Figure 7.5
The block style with subject line is a functional form for letters. Note that the typist's initials are "db"; the recipient will get two enclosures in addition to the letter; and a copy will be sent to another person. The informative subject line provides a helpful and motivating preview of content, and the bulleted list will be used to highlight main points.

relatively good-humored people are irritated when their names are misspelled or their titles garbled. Do not assume goodwill. A female physicist whose first name is Aparoopa may not be amused by a letter addressing her as Mr., while a male engineer whose first name is Robin may give less-than-complete attention to a letter addressing him as Ms.

The Shape of Your Letter

No all-purpose form letter will achieve the results you want for all occasions, for all readers. Like memos, letters must be designed to reach the specific reader named as recipient, the specific readers named as corecipients, and unknown readers who are likely to read the document at some later date.

The recommended three-part organization for memos works well for most letters. Open with an overview, telling the reader exactly why you are writing. Use the middle section of the letter to develop your point. Use the final section to summarize your point and to suggest follow-up action. Use typographical and page design features to highlight key points.

Though the middle sections of technical letters are closely related to the spare and utilitarian style of memos, the openings and closings are more strictly ceremonial. The conventions of openings and closings— "Dear" and "Sincerely yours"—establish a formal tone. Letter writers are even more constrained than memo writers to make verbal gestures that are purely social.

One Subject, One Page

Though memos and letters are frequently many pages long, we recommend using these correspondence forms for brief accounts of single issues, with a goal of one-subject, one-page for each document. The subject should be specified in the subject line, and the content should relate to the stated subject. For two subjects, write two documents. In that way, each subject can receive the full attention of your reader, and each document can be appropriately filed for retrieval at a later date.

Realistically, the conventional format of letters requires so much space for formalities that it is often very difficult to hold to a one-page limit while communicating everything you need to say. Nonetheless, we recommend brevity. If some piece of related information—perhaps a table or figure—is central to the point you are making, treat it as an attachment or an enclosure and refer to it in the body of the memo or letter. Some kinds of reports—trip, incident, progress, and inspection reports, for example—are frequently written as very long memos or letters. An alternative presentation is to prepare the lengthy exposition in report format and to write a separate cover memo or a letter of transmittal explaining the occasion for the attached document and listing major findings.

Page Design Features for Emphasis

For both memos and letters, visual presentation is crucially important: memos look like memos; letters look like letters. But faithfulness to outward appearance is not enough to ensure effective communication. Simply following a prescribed format will not help you to write a memo or letter that suits its particular context.

Though your memo or letter may be brief, do not assume that every word will be read with interest and rapt attention. Ask yourself how you can best design your page for a reader who may not read straight through or who may spend only a minute or so skimming what you have written. Make judicious use of bullets, numbered lists, headings, underscores, italics, and bold type to emphasize the ideas you want to get across. Remember that you are competing for the attention of readers who probably have too much to read and too much to do. The burden of calling attention to key points rests with you, not with your reader.

Electronic Mail

In recent years, the use of electronic mail, or e-mail, to write and post memos and letters has dramatically increased the correspondence workload for many engineers and scientists. E-mail is a powerful communication tool, one that has blurred distinctions between traditional correspondence forms, opened new communication channels, and changed the way that information flows in many organizations.

E-mail written to a colleague in the next office *looks* exactly like e-mail written to a client on another continent. Gone are the social signals and organizational images communicated through letterhead. E-mail document templates carry no honorific titles or forms of address. You, under your user name, write to someone else with a user name. All user names are more or less the same length, without clues to educational or social status. Most e-mail readers open their own mail—even those who never read hard copy memos or letters until a secretary has opened envelopes and logged in each document. Most e-mail readers answer their own mail—even those who otherwise dictate hard copy text for secretarial transcription.

When e-mail addresses are made public, correspondents tend to overstep conventional boundaries created by organizational hierarchies: 65 employees may write to one supervisor, altering long-held conventions about who writes to whom. In networked university settings, many professors note that students are more willing to ask for help with assignments through e-mail than in face-to-face meetings or by telephone.

Getting Your E-Mail Message Read

Some standards for form and style in memos and letters also apply to e-mail. But there is a major difference between hard copy and electronic communication. While e-mail is a supple instrument for sharing ideas and information with others in a local or more fully distributed network, the volume of e-mail in networked writing environments frequently leads to cognitive overload. As a result, e-mail messages are often just skimmed, not scrutinized carefully. A closely related problem with e-mail is that few readers are willing to read extended online text. Important e-mail is often printed out or followed up by a conventional memo or letter.

If you want your e-mail messages to be read, you will have to consider that the recipient of your message may be receiving dozens of messages along with yours. With most e-mail systems, the person to whom you are writing will receive a list of mail to read, identifying the author and displaying the subject line. Nothing obliges a recipient to retrieve and read what you have sent; in most e-mail systems a user can delete unwanted mail without reading it. Ignoring e-mail is as easy as scanning the return address on an unopened envelope and dropping the entire piece of hard copy mail in the nearest trash basket.

As a writer, you naturally want to increase the likelihood that the person to whom you have written will read your message. Try to alleviate cognitive overload by writing a straightforward, information-dense subject line. Keep your message brief: one screenful for one message. Use page design features like bulleted and numbered lists, as you would in hard copy (Figure 7.6). Achieve and maintain credibility: don't send junk e-mail, tempting as it is to take advantage of the ease with which distribution lists can be expanded.

Some e-mail authors are comfortable with more forceful expression (called *flaming*) and more careless style than they would ordinarily use in hard copy memos or letters. When e-mail authors are careless about grammar and spelling, their stylistic informality may be perceived as sloppiness. In corporate settings, where mail goes to many people on large mailing lists and is forwarded and cross-posted, chances are that someone with a low tolerance for grammatical and spelling errors will receive your message. Always assume that verbal restraint and careful editing are valued qualities in professional settings.

Evolving Conventions for E-Mail

E-mail is a technology in cultural transition, appearing to flout much time-honored office, university, and laboratory practice connected with hard copy memos and letters. Much of what happens for both writers and readers of e-mail is constrained or made possible by software design. Most e-mail systems present writers with a template: date and author's name are already filled in; names of others who should receive copies of the message are easy to insert. Even the subject line may be preformed (for example, "Reply to your message of 9/16"). Most templates have no space for anyone's title. You don't need to know whether your recipient has been promoted from Associate Director of Research and Development to Director or whether she prefers being addressed as Professor, Dr., Ms., Mrs., or Miss.

But nothing in electronic communication prevents it from becoming a form with rigid and elaborate social signals. Just as readers of hard copy can quickly size up the importance of a message by noting the

```
Message 1:
From: UNDANK@ZODIAC.BITNET
Received: Wed Jul 29 13:57:46 PDT 1992 from
        mailer@zodiac.bitnet (id AA29665);
To: iwp1muri@UCSBUXA.BITNET
Subject: Lab Manual Revision

I think we should adopt Miguel Barrientos' revision of
the laboratory experiment manual for the Measurement and
Instrumentation course.  He has corrected many of the
flaws in the present manual.  In his version, text is
printed on both sides of each sheet.  This means that
students will be able to complete most tasks for each
section without having to flip pages.  He includes
figures on the same page on which they are discussed in
the text.  And he uses metallic rings instead of staples
for binding the manual, so it is easier to flip through
the pages.  I've asked him to bring it to your office;
let me know what you think.
```

```
Message 1:
From: UNDANK@ZODIAC.BITNET
Received: Wed Jul 29 13:57:46 PDT 1992 from
        mailer@zodiac.bitnet (id AA29665);
To: iwp1muri@UCSBUXA.BITNET
Subject: Lab Manual Revision

I think we should adopt Miguel Barrientos' revision of
the laboratory experiment manual for the Measurement and
Instrumentation course.  He has corrected many of the
flaws in the present manual:

    1.  Text is printed on both sides of each sheet,
        so students can complete most tasks without
        flipping pages.

    2.  Figures are included on the same page on
        which they are discussed in the text.

    3.  The manual is bound with metallic rings
        instead of staples, so it is easier to flip
        through the pages.

I've asked him to bring it to your office; let me know
what you think.
```

Figure 7.6
Compare these two versions of an e-mail message. You can improve the readability of electronic mail by using highlighting techniques like lists and headings to increase readability and emphasize key points.

organizational name and address on the letterhead and the writer's name and title, e-mail templates may be redesigned to provide recipients with social cues to indicate which files can be safely deleted *before* reading and which files need immediate and careful attention. As the volume of e-mail becomes overwhelming, e-mail recipients may create lists of system users from whom they do *not* want to receive communication, and they may ask for unlisted electronic addresses.

The legal status of electronic messages is complex and ambiguous. In 1994, a California court issued a restraining order barring a disgruntled former employee of a software company from contacting the company by e-mail—the first case of its kind. Some organizations are openly monitoring e-mail, and employees have been dismissed for what an employer considered inappropriate or unprofessional comments. Increasingly, e-mail messages, including those assumed to have been erased, are used as evidence in criminal and civil lawsuits. Other cases involving privacy and access are unresolved. E-mail users will do well to write cautiously in this new environment, not mixing the personal and the professional.

Memos and Letters as Part of a Continuum

You can always assume that your memo or letter will not be the last words on a subject. Your document is likely to create additional communication tasks, and its relevance may extend well beyond any time frame you can imagine. Create computer files of memos and letters for future reworking into additional documents. Most e-mail systems provide electronic filing and storing options, though some e-mail users prefer to print and file hard copy of important mail.

Finally, we urge you not to be overly dependent on writing as a method for communicating ideas. Be prepared to *talk* on any subject you have written about. The response to your memo or letter may include telephone calls and face-to-face meetings, both formal and informal. In science and engineering, a written document is hardly ever the only form you will need to communicate about an issue.

8

Proposals

As the manager of a research team, you know that the future of your department depends on a steady stream of funds. You spend part of every month scanning agency releases for possible new grants. So far, you've been awarded enough contracts to sustain your research, but budget constraints have now forced you to limit the size of your staff. No longer can you afford to have a documentation manager available to coordinate, compile, and complete the many proposals that your department submits. You and your staff will have to assume more of the responsibility for proposal writing. And you'll need to find a way to accomplish this task while research continues.

For many engineers and scientists, proposals are the most important form of writing. Most academic research—and a substantial amount of industrial research—is funded through a review procedure in which written proposals are evaluated by panels of researchers from the same field. For working scientists and engineers, proposal writing can make the difference between continued research and interruption in a long-term project.

Proposals set projects in motion and are often part of a cycle of documents that marks the progress of research. They may be preceded by preproposals: a common practice in some industries is to prepare "white papers" describing new concepts and products for likely customers, hoping to receive from the customer a request to provide the items described in the white paper. The work specified in a proposal may be tracked in notebooks and progress reports. Memoranda, reference papers, meeting minutes, and letters then keep a project in motion.

The Great Variety of Proposals

Proposals are written in a variety of informal and formal modes, from short memoranda to multivolume industrial bids. An in-house proposal, written as a brief and informal memorandum, may circulate only within a writer's organization. An external proposal may circulate widely and be refereed by management and budget experts as well as by knowledgeable technical specialists. Depending on the complexity and extent of a research project, a proposal may be written by one or by many researchers and writers. For large industrial proposals, the production group may include, in addition to engineers and scientists, technical managers, editors, text processors, artists, and photographers.

Despite these differences, most proposals have important elements in common. They identify a problem; explain what work will be done to solve the problem; name the researchers who will do the work; argue for their qualifications; specify a time frame, location, materials, and equipment; and calculate a cost. Most proposals are sales documents. Most are planning documents. Most proposals are submitted to reviewers who are knowledgeable, critical, and as quick to reject as to select.

Proposals as Persuasion

A major difference between proposals and other forms of scientific and technical literature is that proposal documents are usually entered into *competitions*. The goal of every proposal writer is to *win* the approval and the money to go ahead with a project. Because success in preparing proposals is a major factor in advancing or even maintaining academic careers, as well as staying in business, writers must overcome any reluctance to draft persuasive documents.

Proposals are mixed bags of elements—technical descriptions, time lines, curricula vitae, budget analyses, fill-in-the-blank data sheets, and more. Think of ways to make every element in a proposal an argument for the value of your idea, the elegance and good sense of your work plan, the strength of your preparation, the appropriateness of your facilities, and the economy of your budget. But a successful proposal requires more than technical details. It requires a story, a narrative shaped to exhibit the strengths of your plan. A well-developed proposal shows that the investigator has grasped a problem well enough to justify second-party sponsorship of the enterprise.

The usual strategy of academic proposal writers is to understate claims, trying to sound somewhat reticent and modest, cautious and competent. In contrast, the usual strategy of industrial proposal writers is to aggrandize. Proposals need just the right sales pitch: the goal is to get a sponsor to spend money. A good proposal must sell, and the key ideas that justify the merits of your approach can be repeated in several places, albeit with modesty and decorum. The industrial attitude, with a frank and forthright interest in winning contracts, may seem overly aggressive to reticent academics. In truth, however, all proposal writers seek to win. Keep that goal in mind, whatever the differences in appropriate decorum.

Proposals as Projections

A proposal is much more than a pitch for financial support. It is a planning document that defines work commitments and establishes the criteria by which the success of a project will be determined. You write a proposal before you know the results. But the illusion that a proposal must foster in its reviewers is that it represents work for which *there is already a plan*. Estimates of a work program, its cost, and its schedule must be convincing. The effective proposal writer must be imaginative, able to convert ideas to tangible projects. You need a touch of the creative writer to spell out detailed plans for 3- to 5-year periods.

Novice proposal writers sometimes come to the task convinced that clever people do not give away secrets until they have won the contract. But a good proposal must describe a project in detail to convince reviewers that they are learning what will happen at every stage of the project. Bidders to the U.S. Air Force, for example, are instructed to be specific: "The proposal shall not merely offer to conduct an investigation in accordance with the technical Statement of Work, but shall outline the actual investigation proposed as specifically as possible." Stating "The work described in the Request for Proposal will be performed as specified" is not much different from saying "Send money."

Of course, a proposal may include a technical design or a management plan that the bidder does not want disclosed. In that case, a Restriction on Disclosure notice, stating that information may not be disclosed for any purpose other than to evaluate the proposal, can be printed on the title page, and every sheet of data that is also so restricted can be marked: "Use or disclosure of proposal data is subject to the restriction on the title page of this proposal."

Proposal Reviews

Reviewers for proposals are usually knowledgeable, critical, and concerned, interested in selecting strong proposals and eliminating problematic ones. Many proposals have multiple reviewers; the more you are asking for—the higher the stakes—the more readers you are likely to have, and the more knowledgeable and critical they will be.

Proposal writers have an important factor in their favor: agencies and other sponsors *need* good project proposals. If research funding is available, referee teams must award their support to someone. Referees want, most of all, to be vindicated in their choices. They have a stake in awarding funds to promising projects. They need more than good concepts; they need evidence that you can meet claims and deadlines. To get well-conceived and well-written documents, agencies often provide extensive criticism to proposal writers.

Strategic Planning for Funding Success

Solicited or Unsolicited

Possibly the most important factor influencing your decision about bidding and the kind of response you prepare is whether the work you propose has been explicitly *solicited* by a sponsor. Proposals are said to be solicited when a sponsor formally announces that funding is available to conduct research in a specific area. Such an announcement may be called a request for proposal (RFP), a request for applications (RFA), or a notice of program interest (NPI). Proposals are considered to be *unsolicited* when they are submitted to an agency that has not circulated a formal request for research. You may also be confronted with a hybrid form: in these situations, the sponsor provides explicit proposal preparation guidelines, but research topics are not specified.

These differences affect the strategy you apply to persuade reviewers to support your project. Solicited proposals must address a problem in an area defined by the sponsor. They will be judged by the writers' ability to meet a specified need, to economize, and to deliver a quality product. These proposals may be measured against their competition on the basis of originality and importance within a discipline. Completely unsolicited proposals present the most severe writing challenge. Here you have the twin tasks of persuading potential sponsors that a problem or a need exists and persuading them that yours is the right group to solve the problem or meet the need (Figure 8.1).

The Right Competitions

Because proposal writing is absorbing and often exhausting, you need to enter the right competitions. If a request for a proposal is available, study it carefully. It will provide detailed information about the work requested and the document required to support your petition. A proposal has the highest chance of success when it is well matched to an assessment of the sponsor's needs. When an agency has explicitly listed areas in which it wishes to sponsor research, research on alternative topics will probably not be funded. The review process is likely to give the largest number of points to projects that are responsive to the agency's request. Proposal documents are not usually the right vehicles for arguing that an agency *should* be supporting research in extragalactic astronomy if the topics listed are all about geophysics.

Occasionally, proposers are encouraged to include alternative plans for satisfying a sponsor's request. In less restrictive cases, an RFP may say that for equal or even preferred consideration a prospective contractor is not limited to the suggested approaches but that any deviations must be fully substantiated. In other cases—and these are real challenges—you need to comply precisely with the terms and conditions stated in the RFP and also to submit a separate alternative proposal, along with a rationale indicating why the acceptance of the alternative proposal would be more advantageous to the sponsor.

Thinking in Two Time Frames

In the challenging work of proposal preparation, you need to think in two time frames: the time you need to prepare the proposal document

Desalination of Salt Water Using Wind Energy

An Unsolicited Research Proposal

M. S. Manalis
Environmental Studies Program
University of California
Santa Barbara, California 93106
805-893-2968; FAX 805-893-8016

March 15, 1993

Table of Contents

i

Abstract

The goal of this proposal is to carry out a feasibility study of coupling wind energy with desalination from small to large scale applications. In many developing and industrialized nations, food production is drastically inhibited by shortages of fresh water. By focusing on use of wind energy to carry out desalination of salt water, a better ratio of on-peak to off-peak electricity generated from wind turbines may be achieved, thereby increasing revenues from wind farms and providing fresh water for drinking, industrial processes, and possible use in agriculture. In many parts of California the wind power density is over 500 watts per square meter. A 34 megawatt state-of-the-art wind farm with a capacity factor of 27% operating in such areas will yield about 5000 acre feet of fresh water annually from desalination of ocean water.

ii

Introduction

Lack of adequate amounts of fresh water is preventing many developing countries from achieving sustained self-sufficiency. The developed countries are consuming fresh water at an ever increasing rate. America consumes about 360 billion gallons of potable water each day (Warfel et al., 1988), approximately three times the rate of consumption 30 years ago. This increase has resulted in the severe depletion of aquifers in the midwestern and southeastern United States...

The purpose of this investigation is to determine the feasibility of utilizing wind energy to desalinate salt water from several sources, such as brackish wells, beach wells, agricultural drainages, and the ocean. California will be the focus of the case study. At present 17,000 wind turbines are producing about 1% of California's electricity. Unfortunately, about seventy five percent of this electricity is produced off-peak. This is a particularly serious problem because peak demand for electricity in California is increasing in relation to base demand. This jeopardizes the future value of electricity generated from wind farms.

Electricity is a perishable commodity; it must be used or stored. Storage of electricity is expensive. When wind energy is used to desalinate water, the final product is not...

1

Figure 8.1

The author of this unsolicited research proposal uses the abstract and introduction to make a strong argument for the significance of the work he intends to

Related Studies

Studies coupling wind energy and desalination have recently been conducted in Europe and America (Carta González et al., 1990; Warfel et al., 1988; Peterson, 1981). These studies indicate that when energy costs are high, wind driven reverse osmosis for desalination probably will be feasible. Cadwallader et al. (1977) reported that coupling of wind energy turbines to desalination equipment offers technological advantages which could be used to optimize the desalination process.

Recent research data in California regarding wind energy and desalination could not be located. However, water planners in Santa Barbara County considered using wind energy on Point Arguello for desalination of sea water (Stubchaer, 1990). Manalis (1979) investigated a wind-driven vapor compressor for use in desalination of ocean water.

Methods

A case study approach will be used with primary emphasis on assessing the potential for developing wind energy desalination in California. The study will...

2

Objectives

This study will encompass the following broad objectives:

1. Analyze interface technology between desalination equipment and wind turbines.
2. Estimate the economic and technological costs of using off-peak wind farm electrical output to desalinate salt water.
3. Evaluate the advantages and disadvatages of using wind energy on site to desalinate salt water.
4. Assess electricity-producing wind turbines as pollution offsets for desalination by conventional energy.
5. Describe the use of wind energy desalination plants to recharge groundwater aquifers.
6. Determine economic and engineering feasibility.
7. Recommend sites for pilot plant if conclusions of study warrant such a recommendation.

3

Project Schedule

This study will be conducted over a twelve month period starting July 1, 1993 and concluding June 30, 1994, according to the following schedule:

Month	Objective
	1 2 3 4 5 6 7
1	
2	
3	
4	
5	
6	
7	
8	
9	
10	
11	
12	
Final Report	

4

References

Cadwallader, E.A., W.R. Williamson and J.E. Westberg (1977). The application of wind energy systems to desalination. U.S. Department of Commerce, Washington D.C.

Carta Gonzalez, J.A. and R. Calero Perez (1990). Optimization technical-economic of the desalination system worked by sind energy. European Community Wind Energy Conference, Madrid, Spain.

Manalis, M.S. (1979). Marine sciences and ocean policy symposium. University of California, Santa Barbara, p. 294.

Peterson, G. (1981). Wind and solar powered reverse osmosis units-design, start-up, operating experience. *Desalination* 39: 125.

Stubchaer, James (June, 1990). Private communication. Water Management Consultant. Santa Barbara, CA.

Warfel, C.G., J.F. Manwell, and J.G. McGowan (1988). Techno-economic study of autonomous wind driven reverse osmosis desalination systems. *Solar and Wind Technology* 5: 549-561.

5

do. He develops a clear list of objectives, and creates a graphic (project schedule) to represent work plans. (Excerpted with permission from Dr. M. Manalis, Santa Barbara, Calif.)

and the time you need to complete the proposed research. In both cases, you match what you have to the sponsor's request.

For the first time frame, some competitions have no fixed deadlines, and some do. For example, most National Science Foundation (NSF) grants for research in education and engineering may be submitted at any time, though some NSF programs set target dates or deadlines for submissions to allow for their consideration by specially assembled review panels that meet periodically. If you know that you cannot get a strong proposal in on time, it may be best not to enter a competition.

For the second time frame, consider whether the research you are proposing is well timed for the announced term of the grant. Funding agencies like results. They will want to be convinced that the work you propose to do in 12 months is actually achievable in that period. Proposal reviewers are usually knowledgeable, and they will prefer funding a summer research project that can be completed in 2 months to funding a summer project that obviously will take 2 years. A graduate student requesting 1 month of research support to review the literature on solid, liquid, and hybrid rocket motors will appear to have a good sense of the work to be completed and the length of the task. An application for a full year of support to do such a literature review will not be very persuasive. Research proposals with long time frames present particular challenges. They need powerful evidence that the work you propose both requires (for example) 36 months of effort *and* can be completed in 36 months.

Read RFPs carefully, looking for information about preferred time frames. Improve your chances of success by proposing projects that fit the agency's guidelines about time. Consider, too, that many agencies acknowledge that research plans have natural phases or breakpoints, and they allow for proposing specific phases of projects. Progress reports are normally expected at the end of such phases, and follow-on funding for additional phases may be available.

Taking Advantage of Help

Overcome any reluctance to take full advantage of assistance. Many agencies encourage you to contact program personnel before preparing your proposal. A meeting will help potential bidders determine if preparation of a formal submission is appropriate. Even if the agency discourages you from proceeding, the feedback you receive may help you develop subsequent proposals. Talk to colleagues who have dealt with the agency or sponsor in the past. Review funded proposals. This is not a do-it-yourself task. Successful proposals require negotiation: you have an idea; you call an agency to discuss your idea; you revise your idea. With every move you narrow the gap between what the agency wants and what you have to offer.

If you are entering a new research area, conduct a literature search on the topic to get a better grasp of any published findings, relevant methodologies, and possible collaborators or competitors. Be sure to consider any political significance of your project. If your research bears on the domain of another scientist or engineer, consider ways to diffuse potential clashes. You might, in some cases, frankly disclose your interests and involve the other researcher at the planning stage.

Evaluation Criteria as Planning Tools

When a sponsor provides criteria for evaluation of proposals, study them carefully at the planning, drafting, and editing stages. To deal effectively with reviewers, you must continually consider *their* constraints and requirements. If criteria for awards are not published in the RFP, it is sometimes possible to get more information by telephoning the funding organization.

Some agencies list areas of evaluation without announcing the weight given to each one. Other agencies provide breakdowns of review categories, so that you can consider the sponsor's weighting as you decide whether to bid. Bidders to the DOT, for example, learn that an unusually sound management approach could offset the problem of insufficient university support (Figure 8.2).

Learning about the Review Process

Proposal funding is not entirely rational. One cynical view of the awards process in science and engineering is expressed in Robert K. Merton's Matthew Principle. Merton refers to Matthew 25.29: "For unto every one that hath shall be given, and he shall have abundance: but from him that hath not shall be taken away even that which he hath," implying that eminent scientists (male *and* female, we will add) may be more likely

Proposal Evaluation Criteria

U.S. Department of Transportation
Annual Solicitation for University Researchers

• Merit of the Technical Approach (possible 40 points)

• Merit of the Management Approach (possible 30 points)

• Qualifications of Proposers (possible 20 points)

• University Support (possible 10 points)

Figure 8.2
In the weighted criteria announced by the U.S. Department of Transportation, a strong management approach could compensate for insufficient university support.

to be funded than less well-known researchers. An explanation of the Matthew Principle, of course, is that eminent scientists submit many proposals and are not deterred by individual rejection.

The effectiveness of the peer review system for evaluating research proposals is a matter of vigorous debate. Critics of peer reviewing argue that reviewers are biased in favor of proposers at the more prestigious universities and that real conflicts of interest can arise between reviewers and applicants. In this view, the system encourages researchers to write devious proposals, not discussing innovative ideas for fear that disclosure will tip off competitors who review the proposal. Critics of the system regard the protective cover of "blind reviewing" as a myth. They point out that knowledgeable scientists in a given field can often deduce the name of the applicant from the very subject of the proposal—and even if they can't, references indicate the applicant's past research and published papers.

Learn as much as you can about the review process for the proposals you submit. The National Science Foundation announces that all proposals are reviewed by a scientist, engineer, or science educator serving as an NSF program officer, and usually by three to ten other experts in the field represented by the proposal. Proposers to the NSF are invited to suggest names of persons they believe are especially well qualified to review their proposal and also, *giving reasons or not*, to suggest persons they would prefer *not* review the proposal.

Approvals in Advance

Proposals that commit the resources of an institution must be approved by an appropriate administrator before they can be submitted. Many universities have administrative units whose responsibility is the administration of contracts and grants. These offices can provide useful guidance. External approvals may be needed as well: a team of physicists at the University of California cannot, for example, guarantee that research will be carried out at Argonne National Laboratories unless they have obtained permission to do so.

In industrial settings, approvals are equally important. An engineering group employed in an electronics firm cannot submit an external proposal without management consent. A proposal is a legal document as well as a sales and a technical document: it binds a company to the statements it contains, and it must be signed by someone authorized to make such promises.

Systematic Proposal Preparation

The proposal is a written product that sets forth the design of a technical product. These tasks mutually govern each other. The tasks of writing the proposal and those of doing the work are often analogous. Both require a systematic approach. Both require knowledge of logical work units. Both require careful estimation of completion time. Both require allocation of responsibility. You can expect, therefore, that the same project management tools used to monitor the progress of writing the proposal are used later to monitor the work defined in the proposal.

Typically, preparing the research plan and preparing the document are not neatly isolated stages. Many sections of proposals are written *before* calculations have been completed, and tidy schemes for proposal preparation and systematic editing seem almost comic to proposal writers working against deadlines. More than one bottle of correction fluid has been used to repair a proposal document, in an automobile, on the way to the post office to meet a midnight postmark deadline!

Our preparation model applies to many kinds of proposals, from relatively brief unsolicited documents written by one researcher to multivolume solicited documents written by a team of 20 or more and coordinated by a proposal manager. Naturally, not every step will apply to every proposal you write, and you will not always have time to attend to every step. Still, the more you are requesting, the more attention you should give to preparing an outstanding document.

Studying the Request for Proposal

Follow proposal preparation instructions exactly. You must provide what the RFP asks for. You do not want your work to be rejected because it has exceeded explicitly stated page limits or does not contain a budget statement matching provided instructions. Read the RFP more than once—and if your proposal will involve other researchers or writers, be sure that all members of the group read the *entire* RFP, not just their own sections.

The instructions for proposal preparation may include a number of forms to fill out and submit with your document: cover sheets, budget sheets, forms for biographical sketches, checklists, mailing labels, and more. Be sure to use each applicable form, following preparation instructions exactly.

Whatever the proposal, it will usually have a prefatory section with a letter of transmittal, cover page, table of contents, list of figures and tables, and summary; a main section with technical, management, and cost details; and some appended items. Complex industrial proposals are often produced as separate technical, management, and cost volumes. Each volume is then evaluated by a specialist. In preparing your document, take care to comply with requirements *and also* to make your compliance visible with sections, titles, and headings that match instructions in the RFP.

If no instructions for proposal preparation are provided, you may select elements with more freedom than writers who must comply strictly with specified requirements. The most sensible plan, however, is to prepare proposals in rather traditional ways, providing answers to standard questions. In these cases, be sure to provide an informative table of contents to show reviewers how your document is organized. The proposal

PROPOSAL FORMAT

Front Matter

    Letter of Transmittal
    Title Page
    Summary or Abstract
    Table of Contents
    List of Figures and Tables
    Compliance Matrix

Proposal Body

    Technical Section
    Management Section
    Budget

Appendixes

    References
    Supporting Detail

Figure 8.3
A standard format for a formal research proposal. Most proposals should explain what work will be done, who will do it, why they are qualified to do it, where they will do it, and with what time frame, location, and cost they will carry out the plan.

format displayed in Figure 8.3 is a conventional one, covering technical, management, and cost areas.

Requirements as Outline and Compliance Matrix
Use the instructions in the RFP to draft an outline for each section of the proposal. You can approach this task in two ways: one is to follow the exact order and numbering system in the RFP. You must cover all points, even if they don't apply to your project, so include every one in your outline. The other way is to prepare an outline in some alternative style, one that appears to be a good way to argue for the value of your

COMPLIANCE MATRIX

REQUIRED	FULFILLED PAGE(S)
FRONT MATTER	
Letter of Transmittal	
Cover Page	
Abstract	ii
Table of Contents	iii-iv
List of Illustrations	v-vi
Compliance Matrix	vii
TECHNICAL SECTION	
Research Objectives	1-3
Methodology	4-6
Anticipated Results	7
MANAGEMENT SECTION	
Task Breakdown	8-9
Timetable	10
Qualifications of Participants	11-14
Related Experience	15-16
Facilities	17-18
BUDGET SECTION	
Cost	19-20
Justification	21-24

Figure 8.4
A compliance matrix shows proposal evaluators that the document responds to requirements in the request for proposal.

proposal and that also *covers every point mentioned in the RFP*. The second method takes more time than the first, and it is somewhat riskier. Nevertheless, some RFPs contain proposal preparation instructions that you may not wish to follow.

When a sponsor provides proposal preparation instructions, a good practice is to prepare a compliance matrix, whether one is requested or not, to indicate your adherence to specifications and to show evaluators where they can find your response to each item required in the RFP. Compliance matrixes (also known as response indexes) indicate the correspondence between the required sections and your document (Figure 8.4). Because they provide a graphic view of what still needs to be done, compliance matrixes are also useful as task management tools.

Time Management
Leave time for your proposal efforts. Industrial bidders to the U.S. government read the government publication *Commerce Business Daily* for notice of upcoming solicitations. In a reasonably typical scenario, advance notice of a solicitation will appear on March 6, the RFP will reach the potential bidder on September 25, and the completed proposal will be due on November 1. Academic proposal writers receive mailings from potential sponsors, so they can plan proposal responses many months in advance.

Nevertheless, proposals are usually written under pressure, when you have technical work as well as proposal writing to do. Often in a matter of 1 or 2 weeks, a project concept must be refined, a team assembled, and a detailed document prepared. This complex process must be thoughtfully sequenced and coordinated to prevent a waste of resources. It must include steps that the proposal group may actually enjoy doing, like technical brainstorming, and steps that many group members will resent, like leaving valuable time for proofing, printing, binding, and delivery.

When you work out a routine for proposal writing, you need to allocate time for each step by first identifying the submission deadline and then backing up to the present (Figure 8.5). The best-made schedules will change. You may need more time than you had anticipated to prepare a budget section and considerably less time to prepare a list of related contracts. Many proposal writers like to use project management software to track their progress; changes can be entered easily, and schedules can be absolutely up-to-date. One software design group at the Sloan School of Management Center for Coordination Science is designing a system that will track tasks, time, and responsibilities. Each night at midnight, the system will check progress and notify project team members who have missed deadlines!

Team Responsibilities
Managing a group preparing a proposal can be as challenging as managing the research itself. In academic settings, the group writing the proposal will probably be the same one slated to carry out the work. Most universities have offices of contracts and grants to give advice at various stages. In industrial settings, many more people are part of

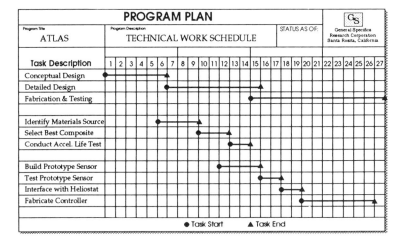

Figure 8.5
Many proposal preparation teams plan and track their tasks with Gantt charts. This group has planned 23 days for writing and 3 days for producing their proposal (*top*). Note, however, that the technical work that is the subject of the proposal goes on at the same time (*bottom*). Few engineers or scientists are able to devote full time to proposal writing!

proposal preparation. In addition to the research group, proposal managers, budget analysts, technical managers, artists, and technical writers are involved. The team's ability to collaborate is a decisive factor in the success of both the proposal document and the funded work.

We think that groups work best when they meet often and when their assignments and responsibilities are visible and explicit. One person should agree to be proposal manager. Every group member should read the entire RFP. Every member should receive an annotated proposal outline with specific allocation of responsibility. Every member should know who is responsible for each part of the proposal. Annotated calendars, printouts of graphics charting project progress, and the compliance matrix should be displayed in prominent places. The group should establish regular meeting schedules, and members should receive explicit instructions about preferred format, writing, and design strategies.

In some industrial settings, the storyboarding method is used to manage collaboration. The proposal manager prepares an outline to match requirements in the RFP, and each member of the writing group takes specific portions of the outline. Authors receive preprinted forms, each representing a two-page spread in the final proposal. They fill in the left side of the storyboard with a thesis sentence and notes about the point to be made in response to their section of the outline. They fill in the right side of the form with rough drawings of illustrations to support the point as well as captions for the illustrations (Figure 8.6). After pinning their storyboards to the wall of a large room, team members can review the document as they walk and talk their way around the room (Figure 8.7).

Storyboarding is helpful because it facilitates revision. Each two-page module can easily be improved without changes to the rest of the document. The process also facilitates review as it makes inconsistencies obvious. It coerces writing that is responsive to the requirements of an RFP, and it makes effort (or lack of effort) visible: blank spaces will show where a delinquent engineer's storyboards should be. In addition, storyboarding produces an efficient document design: tables and figures are always located on the right facing page, directly across from the text passage in which they are discussed. Yet some proposal writers find the method overstructured. They resent the obligation to illustrate. To be

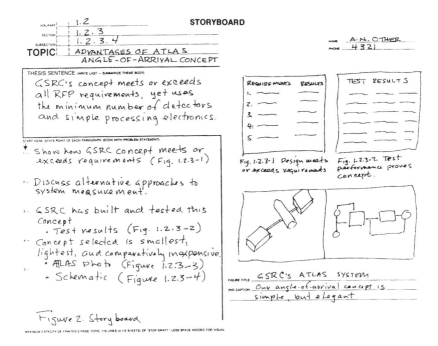

Figure 8.6
This storyboard is a draft of text (left side) and graphics (right side) for two pages of the Atlas proposal. (Courtesy of Hugh Marsh, UCSB)

successful, storyboarding requires a firm commitment from a proposal manager, because it never just happens.

Writing and Format Guides

The most efficient way to achieve stylistic and format consistency in pro-posals is to prepare writing and design specifications. These guides may be as informal as a single-page handout asking all writers to do three things: (1) use the active voice, (2) put important ideas in the first sentences of paragraphs, and (3) use a hyphen when "strip-mine" is a verb. Or they may be lengthy manuals covering numerous issues such as how to prepare mathematical material; preferred spellings, abbreviations, and acronyms; and grammar, capitalization, and hyphenation. The alternative to establishing guidelines in advance is to establish and apply them at the end, when editing time should be spent imposing consistency in mechanical matters such as renumbering equations.

Figure 8.7
Proposal team members pin their storyboards to the wall, and the entire document is reviewed *before* the final draft stage. (Courtesy of Hugh Marsh, UCSB)

Consistency in format is at least as important as consistency in style. An established format shows proposal evaluators that you have prepared the document with considerable care. Your team may find it helpful to distribute samples of finished pages, with heading styles and sizes highlighted, and samples of completed illustrations. Simple format instructions such as "Use Courier font in 10 point, design all figures to fit either one-half page or a full page, and design all figures to be read vertically" may be all you need to ensure that the proposal looks carefully produced.

With many word-processing programs, format specifications can be stored electronically. You can create templates of basic pages with predefined styles of headings, type sizes, fonts, margins, spacings, indentations, and other features. Routines that manage numbering systems for headings, references, equations, and the like are widely available.

Proposals that Comply

The most important criteria for all proposal writing are the explicit instructions in the RFP. Keep in mind that proposals are likely to be reviewed by readers with widely different interests and levels of technical understanding. Because each reviewer will be interested in different aspects, plan to repeat key ideas and make sections of your proposals as nearly freestanding as possible. While full details of the work are provided in the technical volume, a budget analyst may be reading only the cost section. Be sure to provide at least a capsule of information about key aspects of your project in every volume of the proposal.

Front Matter

Letter of Transmittal   A letter of transmittal (or a memo, in the case of an internal document) should always accompany your proposal. As shown in Figure 8.8, the letter should identify the solicitation you are responding to and give a brief overview of proposal contents.

Cover Page   The cover page (Figure 8.9) should at minimum identify the project, the potential sponsor, and the names of principal and coprincipal investigators. Take advantage of the visibility and prominence of the title to teach reviewers about your idea and to sell its advantages. A proposal title should be brief, informative, and intelligible to a scientifically literate but nonspecialist reader. Acronyms sometimes function to produce well-named technical proposals, but they do not work well when the words they stand for are so complex that readers need constant reminders of their meaning. One MIT student suggested the following cautionary acronym:

Adaptive Control and Recursive Optimization of Negative Interlocking Multiplexers (ACRONIM)

Summary or Abstract   In a proposal summary, briefly describe the problem addressed in the study, the methods used, and the expected results. Summaries are typically one to three pages. Think of the summary as a freestanding document, one that may actually have much wider circulation than the rest of the proposal. Be sure that readers can profit from the summary without reading the main body of the proposal: do not refer to tables or figures that appear elsewhere; define acronyms and avoid abbreviations. Some RFPs ask that the summary be written at a level appropriate for an audience of educated but nonspecialist readers.

December 15, 1998

Dr. C.W. Low
Director of Research
California Electric Company
Santa Rosita, CA 93109

Dear Dr. Low:

Enclosed is our proposal for a Tidal Mini-Hydro Power Plant
feasibility study. The proposal was written in response to your
Solicitation #AN248, and we have prepared the document in exact
compliance with instructions provided in the RFP.

Tidal mini-hydro power plants can become an important
supplemental source of energy for many specialized locations.
Because of their potential to alleviate future fossil fuel shortages, a
feasibility study of site selection criteria and technical requirements
is both timely and worthwhile.

We look forward to answering questions you may have and to
working with you.

Yours truly,

Brian Kato
Tidal Mini-Hydro Division

db
Enclosure

2535 West Armadillo • Suite 206 • Santa Rosita • California • 93106

Figure 8.8
A letter of transmittal should identify the competition you are entering and provide a capsule version of proposal contents.

Figure 8.9
The cover page is the most prominent element in a proposal document. An attractive design and informative title can do much to persuade reviewers.

Figure 8.10
Writers of the proposal from which this table of contents is excerpted have used internal headings to argue for the value of their ideas.

In these cases, summaries of successful proposals may be used in agency reports and news releases. Some RFPs ask for "executive summaries," a term you should read as an instruction to write for nonspecialist readers.

Like a summary, an abstract may have a life of its own and be read by far more readers than those who evaluate the proposal. Abstracts are typically briefer than summaries (one paragraph of approximately 150 words), and they are written for the same specialist readers who will read the proposal (see Figure 8.1). Some RFPs ask for both summary and abstract.

Table of Contents   In the table of contents, list primary and secondary subject headings and descriptions of what is contained in appendixes. Provide a page number for each element. Edit the table of contents for consistency and persuasiveness: because it is so prominent a section, it can function as an important sales pitch for your idea. The headings excerpted from the proposal in Figure 8.10 are explicitly written to help evaluators reach positive conclusions about the work described. Instead of writing perfunctory headings like "Project Placement," the proposal team has written headings like "Project Placement Takes Maximum Advantage of the Expertise of Three GSRC Groups."

List of Figures and Tables   List all figure and table titles and their page numbers. The list of figures and tables is highly visible and widely used by technical readers. As with headings, you can use titles to inform and persuade. Instead of writing perfunctory titles like "Filtering system," you can write titles that lead evaluators to the conclusion you hope they will reach: "Filtering system has been modified to exceed requirements."

Compliance Matrix   Whether or not one is called for in the RFP, a compliance matrix (see Figure 8.4) indicates that you have paid careful attention to the sponsor's requirements. It also tells reviewers where they can find your response to each required section.

Body of Proposal

Technical Section   In the technical section, identify the problem and its significance, state the objectives of the proposed investigation, and review previous work and related studies. Outline your general approach to the research. Note significant alternatives and your reasons for not pursuing them.

Management Section   The management section names the personnel who will do the proposed work and the facilities in which the work will be done. It also contains highly detailed task breakdowns and work schedules. Management sections are the place to argue for the qualifications of the principal investigators and their associates. Relevant highlights of curricula vitae can be summarized, and lists of previous related contracts can be provided.

Budget   In your budget, provide cost details for salary and benefits, and justify each number. Include indirect costs (overhead), as well as direct costs like materials, equipment, salaries, and travel. Many sponsors provide preprinted sheets for budget calculations, and you must use them to comply with the RFP.

Appendixes

References   List references to previous papers, documents, and discussions that have been used in preparing the proposal.

GSRC Understanding of the Problem

General Specifics Research Corporation Santa Rosita, California

Tidal Mini-Hydro Power Plant

Problem	Approach
Corrosion	Induced Reversed Voltage
Water Quality	Incorporate Trash Racks
Energy Storage	Batteries and Flywheels
Vortexing	Raft/Grid to Eliminate Vortices

Figure 8.11
Skillful proposal writers use every opportunity to convince reviewers that they have solutions to problems. In this graphic from the executive summary of an industrial proposal, reviewers are told that the bidders understand the task.

Supporting Details When appropriate, include items like curricula vitae, copies of publications of principal investigators in areas related to the proposal topic, lists of previous related contracts, letters of reference, and detailed and oversized figures and tables.

Stressing the Strengths of Your Ideas

Because a proposal is a sales document, you need to be able to identify and emphasize the features and benefits of your idea. At the paragraph and sentence level, in section previews, in specially designed graphics, and in legends to illustrations, skillful proposal writers can reinforce powerful arguments for the value of their plan (Figure 8.11). Lead with your strengths, putting important ideas in prominent places.

Give reviewers the impression that you are confident of success by using the present tense for general descriptions and the future tense for actions in the future. Write as though the funds and approvals *have been granted*. A proposal style dependent on conditional verb forms is awkward ("If funding were to be granted, we would at the time of the second phase of the project develop test equipment to support the ATLAS technical initiative"). It indicates lack of self-confidence, though the writers may simply have intended to be modest and tactful until they actually receive the award. Compare "In Phase 2 we will develop test equipment."

In some particularly competitive industrial settings, proposal groups create a "win theme matrix." They list strong thematic phrases or sentences that argue for the merits of their idea—and they write these phrases into every section of the proposal, using the matrix as a checklist to assure that they have repeated their win themes in every section.

Preparing an Attractive Document Package

Though proposal evaluation schemes never award any points for visual attractiveness, the way a document looks can convey a powerful sense of your professionalism and competence. Even when a proposal is written in accord with rigid and challenging page limits, its design elements can facilitate navigation through the document. Tabbed section dividers and informative page headers help busy reviewers read your document efficiently. Judicious selection of type styles and sizes will signal what elements are more important than others. A heavyweight or even laminated proposal cover may assure that your document will hold up to multiple reviews. A logo, proposal title, and organizational name on *every* page will serve as a reminder to reviewers of who you are and what you are selling! Successful proposals—both industrial and academic—are often exceptionally attractive documents, with numerous foldouts, photographs, and other artwork set on pages designed with great care.

But in selecting design options, be sure to read the RFP for guidelines. The National Science Foundation could hardly be more explicit in its call for utilitarian design and packaging. The *Application Guide to Grants for Research in Science and Education* (NSF 90-77) states: "Proposals should be stapled in the upper left-hand corner, but otherwise unbound, with pages numbered at the bottom and 1-inch (2 cm) margins at the top, bottom and on each side, in type no smaller than 12 point. The original signed copy should be printed only on one side of each sheet."

Your Proposal Writing Program

Resubmitting

It is highly unlikely that every proposal you write will be funded, and it is hard to know exactly why one proposal wins and another does not. If your project is denied, you can still get some benefit from the work you have done by finding out why, a process called "getting debriefed." The amount of information you can receive about the deliberation process varies from agency to agency, although the identity of reviewers is never revealed.

If you have submitted your proposal to the National Science Foundation, you will automatically receive such a debriefing. When the decision is made, the principal investigator receives copies of reviews (excluding the names of reviewers), summaries of review panel deliberations, a description of the process by which the proposal was reviewed, and details about the decision, such as number of proposals and awards. Other agencies will provide debriefings with a written request, usually within 30 days after the announcement of the final selections.

More extensive information is sometimes available. Investigators whose proposal for NSF support has been declined may request additional information from the program officer or division director. If that official's explanation does not satisfy the investigators, they may request a reconsideration to determine whether the proposal received a review that was fair and reasonable.

Depending on what you learn in your debriefing, you may want to resubmit the same proposal or one substantially like it in a future competition. Possibly it was technically excellent but did not fit the agency's research priorities for that year. If you identify weaknesses in your project proposal, consider ways of salvaging the concept. In some instances, you may focus the proposal on another area and apply to another agency. The NSF specifies that a declined proposal may be resubmitted only after it has undergone revision. At that time, the NSF will treat the revised proposal as a new one subject to the usual review procedures.

Document Databases

If you regularly write proposals, you will want to create computer files of standard information, text, *and* graphics. Many proposal sections—including curricula vitae, drug-free workplace plans, related experience, and management history—are likely to be required in nearly the same form for any project you may bid on. Instead of compiling and typing these chunks of standard text each time a proposal is created, you can record and save them as separate files that can be quickly tailored and inserted into new documents as needed (Figure 8.12).

You will, of course, want to review these chunks of boilerplate as you use them, but they are easily updated to contain the most relevant text and graphics. An additional benefit is that these data files can be made available for use by other proposal writers in the organization.

Staying Informed

A proposal-writing program requires careful planning. You must know the needs of your discipline; you must have good information on funding sources and their requirements. Keep a file of agency announcements, RFPs, and NPIs. Keep another file of new project ideas that occur to you or to members of your team. Articles from the literature may suggest new research possibilities for your field. To develop new ideas, go to conferences, read the *Annual Register of Grant Support*, and keep up with the literature. Look for technological developments that make new research feasible; some researchers sift through patent literature for ideas. The FEDRIP database of *fed*eral *r*esearch *in p*rogress can be accessed at many technical libraries. The National Science Foundation home page (http://stis.nsf.gov/) is a good source for proposal announcements and deadlines.

If you depend on the support of sponsored research, you must have a long-range funding strategy. If you are writing proposals in June for next winter's support, you're in trouble. Most research proposals take from 4 to 9 months for review, and you may need to be thinking 2 to 3 years ahead of the present. Track new project possibilities and funding sources. Keep agency application deadlines prominent on your work calendar, and meet each annual application deadline with one or more new proposals. A professional researcher may have 5 to 10 proposals circulating

TYPICAL AMBER ENGINEERING PROPOSAL STRUCTURE

Section Number	Description	Text is Usually
-	Executive Summary	New
1	Introduction	New
2	Technical Approach	New
3	CMOS Technology	Standard
4	Test Approach	Modified
5	Related Experience	Standard
6	Program Plan	Modified
7	Proprietary Technology	Standard
8	Statement of Work	Modified
9	Organization	Standard
10	Resumes	Standard

1

STEPS TO IMPLEMENTING THE USE OF GLOSSARIES AT AMBER

• Identify Recurring Proposal Content

• Create Glossary Entries

• Make Glossaries Available for Use

• Educate Proposal Managers and

Other Interested Parties

• Maintain Current Glossary Data

• Continue Improving Proposal

Preparation Methods

2

Figure 8.12
In this presentation to management at Amber Engineering (Santa Barbara, Calif.), engineer Stan Laband argues for the creation of document databases. Of 11 typical proposal sections, 5 are standard in all proposals, 3 are modified, and only 3 are new. Creating and reusing stored text will yield improved accuracy in content and consistency in format. (Courtesy of Stan Laband.)

at once. To write proposals on this scale, you obviously need to work out a detailed application routine.

Proposals as Part of a Continuum

Proposals are crucial documents in the production of scientific knowledge, providing access to the funding and approvals that make research possible. When proposals are successful, they create more communication tasks. They lead to writing projects such as progress reports, final reports, conference proceedings, and journal articles. Each document form disseminates information to wider and wider audiences.

Proposals also create occasions for *speaking* about your ideas. When you work on a proposal, be prepared to talk persuasively about the project on occasions ranging from informal telephone conversations and hallway meetings to formal presentations with potential sponsors. In active professional settings, you will not always receive advance notice of the need to speak cogently about your ideas, but do assume that you should always be prepared to talk about your research. Like a written progress report, an oral summary is a small but significant way to argue for your position.

9

Progress Reports

You're in a bind. Your third progress report is due tomorrow, and the news is bad. Less than halfway into a 12-month project, you've fallen behind schedule and used close to half your budget. You still hope to compensate for lost time and extra expenses, but now you must tell your manager about these difficulties. Unfortunately, you've already compounded your problem. Last month, you avoided reporting the project's deficiencies, hoping that you could solve the problems quickly. Your manager will be unpleasantly surprised that your last report was overly optimistic.

Progress reports are always intermediary, never final documents. They track, evaluate, and archive the work of science and engineering. Progress reports are bridges, spanning the time between the beginning and the end of a project—from the projections in the proposal to the reality of the work.

In the proposal that initiates a project, you describe tasks and activities that have not yet happened as though you know what will happen at every stage of the work. You estimate, guess, project. Your proposal has minimized ambiguity. Instead, you have described a predictable series of events leading to a successful outcome.

In contrast, you write a progress report as you do the work. Your report might tell the same story as the proposal, or it might tell a different one. A proposal says, "This is what will happen"; a progress report says, "This is what actually happened." The baseline against which progress is measured is what you said would happen: at what cost, in what time frame, with what researchers involved, and with what deliverables delivered.

The Great Variety of Progress Reports

Progress reports may be formal documents written to satisfy external funding sources, or they may be informal documents that track and monitor employee activity within an organization. In many industrial settings the progress report (sometimes called a status report, a morning report, a briefing, or a weekly) is one of the most frequent and routine pieces of writing for technical professionals. Typically, each researcher prepares an account of activities for a specified period and passes the account upward to a manager. The manager writes a summary progress report for an entire department and passes this more comprehensive account upward, to a manager at a higher level. At the highest level, a manager will be able to report on progress for numerous projects.

Audiences
Readers of progress reports are likely to be knowledgeable about the technical areas you describe and also deeply concerned with the status of activities. They want to know what has been done and what needs to be done, what problems you have encountered, and how likely you are to stay within a previously agreed-upon budget and schedule. They want to know if you are spending their money and your time in ways that will yield desired outcomes.

Because such readers are typically worried about the project, writers sometimes feel constrained to use the progress report to allay anxieties rather than record problems. But the progress report *is* the vehicle for assessment, evaluation, and possibly for renegotiation. Your report may cause the project to take a new direction, with revised goals. You might be tempted to use the progress report for claiming credit, showboating your achievements, crowing about the excellent match between your projections and what has actually happened. Don't. It is not only more decorous but also safer to report achievements modestly and to report problems as soon as possible. Progress reports have been entered into legal proceedings. You are on solid ground if you record what has happened, not what you want the customer to think has happened. Recognize that authors of progress reports are not always able to record desirable advances toward the orderly achievement of a goal.

Your progress report should address the concerns of your readers by including a section devoted to assessment and evaluation. A genuinely useful progress report will include interpretation of results as well as raw numbers.

Formats and Schedules

In writing a proposal, you follow instructions provided in the request for proposal (RFP), or you create a responsive format of your own. In the same way, a progress report follows instructions provided by your sponsor or your supervisor, using the forms and formats specified, or you create your own functional document. For many government contracts, the frequency of progress reports will be specified in a contract data requirements list (CDRL), with details of form, length, and style provided in a data item description (DID).

Progress reports vary widely in form and length, from a one-page memo, letter, or preprinted form to a several-hundred-page bound volume with a separately bound appendix. In tracking a multiyear project, progress reports are often long at some stages, perhaps annually, and very brief at others.

Progress reports tend to be less formal than proposals or research reports. Each one summarizes progress since the preceding report, and the earlier report is always summarized, subsumed, and superseded by a new one. Progress reports have some elements in common with proposals and formal reports, but they are always distinguished by their focus on:

Project status
Measurement of achievements against projections
Problems encountered
Work completed
Work remaining
Evaluation

Figure 9.1 lists possible elements in a progress report. In selecting elements, remember that all progress reports in a series should have the

PROGRESS REPORT FORMAT

Front Matter

    Letter of Transmittal
    Cover Page with Date and Number
       in Series
    Table of Contents
    List of Figures and Tables
    Project Summary

Report Body

    Work Accomplished to Date
    Work Remaining
    Plans for Next Reporting Period
    Problems Encountered
    Appraisal of Progress to Date
    Recommendations

Appendixes

    References
    Supporting Details

Figure 9.1
Possible elements in a formal progress report. If you need to prepare a series of progress reports for a single project, use the same elements and formats for each.

same format. For example, if you have included a section called "Problems Encountered" in the first report, you should include the section in the second report, even if all you have to say is that you have encountered no new problems.

The schedule for submitting progress reports is established in a proposal (Figure 9.2) or by organizational practice. Reports may be required weekly, monthly, bimonthly, annually, or on some other schedule. Often, one kind of report is required weekly while another kind is required semiannually. Whatever the schedule, you must follow it strictly, even if your work has not gone as you had planned.

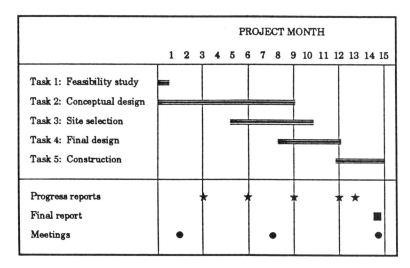

Figure 9.2
Five progress reports are scheduled during the 15-month period of this project. The schedule was established in the proposal.

Organization

Readers review progress reports to get answers to *their* questions about project status: perhaps monitoring employee productivity, perhaps keeping an eye on costs. A well-designed progress report will be modular, with modules clearly labeled, so that readers can skip over sections they do not need to read and proceed quickly to sections that interest them. No reader will want to spend time hunting for information. Use headings for each new topic area, and provide a table of contents for progress reports of five pages or more. Consider using labeled tabs to divide and mark sections in longer reports.

Figure 9.3 illustrates alternative ways to present information in a progress report. Progress report 1, the more conventional of the two, is organized chronologically, with the emphasis on the current state of the project. In this version, the tasks are subdivisions. Progress report 2 is organized by task, and in this version the current state of the project is subordinated. If a sponsor has not given explicit instructions for preparing a progress report, you can choose a presentation style that best reflects the work. Style in the narrative sections of progress reports is sometimes relatively formal, resembling the style of a proposal or final

Progress Report 1

Work accomplished to date:
    Task One: Evaluate Site
    Task Two: Treat Existing Coal Pile
    Task Three: Install Storage Pile Sprinklers
    Task Four: Install On-Site Testing Facility

Work Remaining:
    Task One: Evaluate Site
    Task Two: Treat Existing Coal Pile
    Task Three: Install Storage Pile Sprinklers
    Task Four: Install On-Site Testing Facility

Plans for next reporting period:
    Task One: Evaluate Site
    Task Two: Treat Existing Coal Pile
    Task Three: Install Storage Pile Sprinklers
    Task Four: Install On-Site Testing Facility

Progress Report 2

Task One: Evaluate Site
    Work accomplished to date
    Work remaining
    Plans for next reporting period

Task Two: Treat Existing Coal Pile
    Work accomplished to date
    Work remaining
    Plans for next reporting period

Task Three: Install Storage Pile Sprinklers
    Work accomplished to date
    Work remaining
    Plans for next reporting period

Task Four: Install On-Site Testing Facility
    Work accomplished to date
    Work remaining
    Plans for next reporting period

Figure 9.3
Several organizational patterns are appropriate for reporting progress. Progress Report 1 emphasizes the overall status of the project; Progress Report 2 emphasizes the status of individual tasks.

Feasibility Study: Tidal Mini-Hydro Power Plant

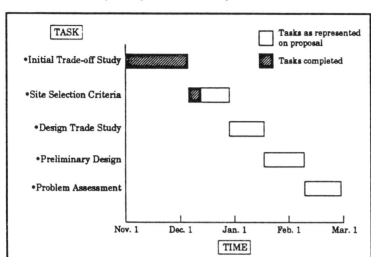

Figure 9.4
Good graphics for progress reports show what was supposed to be accomplished
and what actually was. They serve as planning tools for researchers as well as for
project sponsors.

report, and sometimes informal, with phrases acceptable in place of
complete sentences.

Design and Function

Progress reports often have more graphic than narrative information.
Figure 9.4 displays the tasks defined in the proposal and provides a rapid
view of the status of the project, showing which tasks are completed and
which are not. Establish one style for graphics to chart and track your
progress. Use the same graphic forms for each progress report in a series.

Many widely available software packages can be used to create spreadsheets, tables, and figures, so that monthly or quarterly reports can be
easily compiled by updating the last report. Photographs provide vivid
and dramatic information about progress on some kinds of projects. As
computer workstations more routinely have multimedia capabilities,
requirements for many progress reports are likely to include video and
audio—giving the nervous sponsor, several thousand miles from your
laboratory or construction site, more information about your progress.

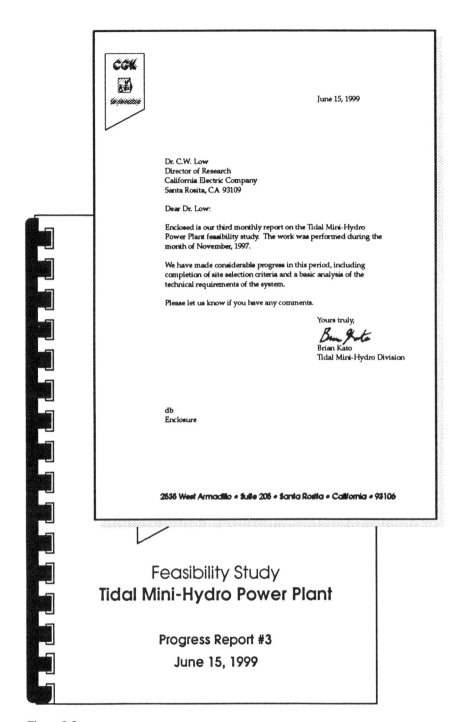

June 15, 1999

Dr. C.W. Low
Director of Research
California Electric Company
Santa Rosita, CA 93109

Dear Dr. Low:

Enclosed is our third monthly report on the Tidal Mini-Hydro
Power Plant feasibility study. The work was performed during the
month of November, 1997.

We have made considerable progress in this period, including
completion of site selection criteria and a basic analysis of the
technical requirements of the system.

Please let us know if you have any comments.

Yours truly,

Brian Kato
Tidal Mini-Hydro Division

db
Enclosure

2635 West Armadillo • Suite 205 • Santa Rosita • California • 93106

Feasibility Study
Tidal Mini-Hydro Power Plant

Progress Report #3
June 15, 1999

Figure 9.5
Progress reports should be prepared not as throwaway documents but as serious
pieces of technical literature. They can be attractively packaged and, for external
sponsors, accompanied by a letter of transmittal.

Though progress reports are intermediary documents, they are not throwaways. Create a functional and attractive package for your report. Progress reports with external audiences should have front and back covers. Some companies provide their sponsors with loose-leaf binders in which to store numerous reports in a series. For progress reports submitted outside your own organization, include a letter of transmittal, just as you would with a proposal or a final report. In Figure 9.5, a progress report is accompanied by a brief letter giving major highlights of the reporting period.

Document Databases

You write progress reports while you are engaged in scientific and engineering work—when you are very busy, not when you are on vacation! You will want to streamline the process, reducing the amount of time you need to spend on each report by adopting a functional document format, establishing a style for project-tracking graphics, and creating computer files of standard information, text, *and* graphics. Some text sections—perhaps a safety report or a weather impact assessment—will be required in nearly the same form in every report in a series. Instead of typing these chunks of standard text each time, you can record and save them as separate entries that can be quickly tailored and inserted into new documents as needed.

A progress report is, by definition, not the last word on anything. But it has a crucial role in the continuum of documents that track the life cycle of a technical project, measuring what happened against what you hoped would happen, providing opportunity for rethinking and negotiation. And a progress report provides content and direction for the final report, which will mark the conclusion of a project. When you have finished a series of progress reports, you may find that you have already written a substantial part of the final report.

10

Reports

As part of a team busily finishing a funded research project on a tight deadline, you know that the final report will require some careful planning to be completed on time. You've discussed the report with other team members, but so far no one seems willing to consider its scope or outline. One colleague has assured you that a final report could be written by compiling parts of the series of progress reports the team has submitted. Your task, then, is to convince team members that they must begin work on the final report soon. If they put it off much longer, the deadline will be in jeopardy.

In science and engineering literature, the term *report* describes a document that tells what happened. A report may record experimental results, assess project feasibility, make recommendations for action, or provide observations and commentary on an inspection trip.

A report may appear within another document, perhaps a memo or a letter, but larger formal reports are prepared in book format, with title page, table of contents, lists of illustrations, sections or chapters, and appendixes. These final reports contain some of the familiar elements of proposals and progress reports, but the emphasis in a formal, final report is always different. A proposal says, "This is what will happen"; a progress report says, "This is what has been happening and what is expected to happen next." A final report says, "This is what happened."

Reports on the Writing Continuum

Just as a progress report is never the last word on a subject, a final report is rarely the first word. The goals and achievements of a technical project have probably been written up in laboratory notebooks, meeting minutes, proposals, or progress reports.

In the continuum of written work that proposes and records scientific and engineering activity, some elements are unchanged from document to document. A literature review, for example, might be essentially the same in the proposal and the final report. Some elements reflect only a change in emphasis: a final report may account for the time management of a project, but it is unlikely to focus on this area as much as a proposal must. In a final report, some elements are eliminated: a final report rarely contains curricula vitae of investigators, for example, but a proposal usually does.

Availability of Reports

In most cases, a proposal has very limited circulation, while the availability of progress reports that track the status of the funded work is somewhat larger but still limited. Completion reports, in contrast, may be widely disseminated, though never so widely as journal articles. Technical reports originating in private industry are usually proprietary documents, circulated internally and not available to outsiders. But reports that result from federal grants will be indexed in the National Technical Information Service (NTIS) database, or *Government Reports and Announcements Index*. Additional on-line and printed reference services that provide access to reports include *Engineering Index*, *Chemical Abstracts*, and the National Aeronautics and Space Administration's STAR index. In current practice, reports are made accessible to other interested researchers though electronic searches of titles, abstracts, and keywords. With increased use of electronic publishing, researchers can sometimes do full-document searches. Therefore, the limited circulation accorded a proposal document contrasts significantly with the potentially vast audience for a *nonclassified* final report, which becomes part of the permanent record of what is known on a subject.

Report-Writing Conventions

Report writing takes a good deal more intellectual activity than following a formula or a recipe. In writing some reports, you will be

TECHNICAL REPORT FORMAT

Front Matter

Letter of Transmittal
Title Page
Abstract
Table of Contents
List of Figures and Tables
List of Abbreviations of Symbols

Report Body

Introduction
Theory
Experimental Section
Results
Discussion
Conclusion
Recommendations

Appendixes

References
Supporting Details

Figure 10.1
Conventional elements in formal technical reports.

given a prepared outline—an intellectual template—and required to write standard sections. For other reports, you will need to devise a structure that suits your purpose, the needs of your audience, and your subject.

Technical reports usually have a three-part structure: front matter, report body, and appendixes. Within these three broad divisions, reports vary widely in choice of elements and degree of emphasis on any one element. While a laboratory report may emphasize the way an experiment was performed, a management report is more likely to emphasize conclusions and recommendations. Figure 10.1 lists conventional elements for formal reports on scientific and engineering subjects.

Front Matter

Letter of Transmittal   The letter of transmittal accompanies the report, identifies the item being sent, and provides a context. The letter also provides a brief overview of report contents, typically emphasizing findings of general interest (Figure 10.2).

Title Page   The title page should indicate names of authors, date on which the report was submitted, organization or institution in which the report was prepared, report number or other indication of the *occasion* for the report, and proprietary notices if they are appropriate.

A report title should name the subject in as straightforward a way as possible. The title should serve as the report in miniature for the widest possible audience. Construct titles with informative words. In some online databases, a keyword list is electronically generated from your title, so you need to build a title from words that represent the most important concepts in your document.

Titles can often be improved if you eliminate inessential detail. The title "Survey and Evaluation of Electrical Power Sources as to Their Potential Application with the Controlled Airdrop Cargo" is overloaded with words that have little information value. A simpler version cuts the number of words and focuses on content: "Potential Electrical Power Sources for Controlled Airdrop Cargo."

Abstract   Most scientific and technical documents contain an abstract, a concise account of the problem addressed and the results. The format of abstracts sets them off from the rest of the document: they are typically written as a single paragraph and printed single-spaced, indented on both sides.

Abstracts are classified in two types: informative and descriptive. Informative abstracts are typically about 150 words long, and they present methods, results, conclusions, and recommendations of the report in miniature. An informative abstract frequently stands for the entire report; it may be all nonspecialist readers want to know about your research. Descriptive abstracts are often no more than a sentence, and they may not go much beyond the information already presented in a

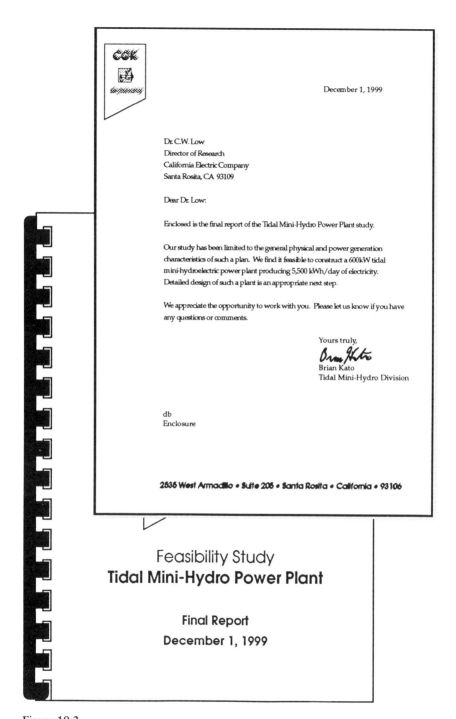

CCK

December 1, 1999

Dr. C.W. Low
Director of Research
California Electric Company
Santa Rosita, CA 93109

Dear Dr. Low:

Enclosed is the final report of the Tidal Mini-Hydro Power Plant study.

Our study has been limited to the general physical and power generation characteristics of such a plan. We find it feasible to construct a 600kW tidal mini-hydroelectric power plant producing 5,500 kWh/day of electricity. Detailed design of such a plant is an appropriate next step.

We appreciate the opportunity to work with you. Please let us know if you have any questions or comments.

Yours truly,

Brian Kato
Tidal Mini-Hydro Division

db
Enclosure

2535 West Armadillo • Suite 205 • Santa Rosita • California • 93106

Feasibility Study
Tidal Mini-Hydro Power Plant

Final Report
December 1, 1999

Figure 10.2
The letter of transmittal accompanying the report provides a context and a summary of contents. It ends by expressing appreciation for the opportunity to do the work.

INFORMATIVE ABSTRACT

Remote Sensing of Automobile Exhaust. P.L. Guenther et al.
American Petroleum Institute Report 841-45380 (1991).

An on-road remote-sensing (up to 40 feet) hydrocarbon channel, added to an existing infrared remote sensor that detects passing automobile CO and CO_2 exhaust emissions, showed that 14 percent of a fleet of 3,000 vehicles in Denver, CO contributed half the propane emissions. The instrument measures CO, CO_2 and hydrocarbon emissions to a precision of 0.15, 0.15 and 0.02 percent of absolute concentration respectively. Interference from water vapor in the exhaust is less than 1,000 ppm propane. The hydrocarbon sensor, with noise of about 150 ppm, is not as precise as conventional nondispersive infrared units (NDIR) but can make more than 1,000 CO and hydrocarbon measurements in an hour, compared with a maximum of about 20/hour for NDIR units.

DESCRIPTIVE ABSTRACT

Fuel Efficiency of Passenger Cars. *International Energy Agency Report* 92-64-13463-8 (1991).

This report examines factors affecting fuel demand in International Energy Agency member countries, describing how regulatory measures to limit pollution by passenger cars have affected fuel use.

Figure 10.3
An informative abstract presents the report in miniature, whereas a descriptive abstract is more like an expanded title.

title (Figure 10.3). Unless you have been given explicit instructions to the contrary, you should prepare an informative abstract for any report.

Abstracts are most profitably drafted after the report has been written. Think of an abstract as a smaller document that describes, but does not evaluate, a larger document. Unlike an executive summary, an abstract does not need to simplify technical concepts or sell a subject. The audience for the abstract is likely to have the same level of technical understanding as the audience for the full report. The tone of an abstract is objective, not persuasive.

Table of Contents   The table of contents is an important section of any report, even a short one. Include primary and secondary headings, thereby giving readers a quick overview of report contents. A table of contents can also help you construct and reconstruct cogent documents. The headings provide a structural view of the document. They may reveal organizational defects that would not be easily obvious in line-by-line reading.

List of Figures and Tables   If you have two or more figures or tables, provide a list with full legends. Technical readers often determine whether a report will be of interest to them by scanning the list of figures and tables (sometimes called list of illustrations). Like tables of contents, these lists are also helpful to you as you revise your writing. A review of the legends in your list of illustrations may reveal defects in logic or organization. Sometimes, you can find better ways to present your findings by relocating illustrations or writing more informative legends.

Lists of Abbreviations and Symbols   Though a list of abbreviations and symbols (also called a glossary) may not be required, you should consider whether your readers could use such help. In many fields, these lists of terms are essential; even active researchers can hardly keep up with the growing number of abbreviations and acronyms. When you provide a glossary or list of abbreviations, however, you still define each term the first time you use it in your report.

Report Body

Introduction   The introduction to a formal report has three functions: (1) to define the problem addressed, frequently with a review of previous work in the area; (2) to state explicitly the objectives of the present work; and (3) to summarize the main conclusions or applications of the work. Introductions may be one paragraph or several pages long. Though problem definition is the key to a successful report, all three functions of the introduction are important.

The move from defining the problem to stating objectives and conclusions is awkward for many writers. Having launched a problem discussion, you may feel uncomfortable turning abruptly to stating just what the objectives of your document are. Graceful or not, a transition must be decisively executed before the paper can begin. Experienced report readers scan introductory sections for sentences with openings like "In this report, we examine...." They know they will find in such sentences a concise statement of topic and findings.

Theoretical Section In the theory section, you discuss criteria for addressing the problem and outline your general approach to the research. Authors of research reports may develop and present their own models, or they may rely on other published work, providing citations to the papers that developed the original models. The solutions for most problems do not require you to develop basic principles but to apply old ones.

When you rely on models developed in other published work, cite the papers that developed the original models and establish why you are citing them. Take care to review the literature accurately and carefully. Simply listing reference numbers at the end of a sentence can raise doubts about your use of sources. Be sure to explain what you are applying from each source.

Experimental Section The experimental section describes the laboratory equivalent of your problem: the tools and processes that enabled you to meet your stated objectives. Here you convert your concept of the problem into the language of the laboratory, so that readers may test your methodology against your results in their own laboratories. Clarity and accuracy are priorities. You are describing a variety of objects, materials, processes, and instruments that, used in a specific way, must deliver a characteristic set of data.

In all technical description, an overview helps readers grasp the purpose and scope of your experimental work. Complex procedures are more effectively described when they are arranged in discrete subsections. Consider providing a drawing for any complex apparatus you intend to describe in detail. And for complicated preparation processes or experimental procedures, consider placing the detail in an appendix.

Results   The results section translates your findings into the terms of closely observed phenomena, the data of instruments, and the language of numerical generalization and statistical analysis. All the discussions of your report either lead up to or away from this often brief section. Results sections often retain their value long after the methods and conclusions have become obsolete.

Data may be presented in many ways. If the results are simple, note them in a prose passage. Present series of data in tables or graphs. Avoid simply noting that the data are shown in a given table or figure. Rather, draw out an important point or two about the trends shown in each table or graph and emphasize these points again in titles or legends. If your methods of reducing data or estimating their accuracy are based on other published sources, provide the reference.

Discussion   The purpose of the discussion is to evaluate results. Results do not speak for themselves; you must interpret them. Interpretation should return the reader to the original objectives stated in the introduction, as well as to the initial theoretical discussion.

Begin by amplifying the most important findings and noting any significant discrepancies. Most authors also discuss the reliability of their main findings. If you can identify errors in your own work, you lend credibility to your discussion by noting them. Even when there is no clear explanation for inconsistencies or errors, you should note their existence. To develop a discussion more fully, you may compare your results with those of other sources.

Conclusion   The conclusion section restates main findings, summarizes results in light of the original problem, and draws generalizations supported by those results. The conclusion may contain suggested applications for research results, or it may connect results to other scientific issues. Sometimes the conclusion is combined with the recommendations section.

Recommendations   In reports written for management, recommendations sections occupy a prominent place, while in technical reports, recommendations are usually a final, brief section, sometimes combined with the conclusion. You may or may not wish to make recommendations about directions that future research on your topic should take. If the main objective of your report has been to recommend a specific course of action, you will naturally devote an entire section to outlining a concrete and operational set of moves.

Appendixes

The appendix (plural: appendixes) contains information that is so excessively specialized, so lengthy, or so unwieldy that placing it in the body of the text would interfere with reading. Appendixes frequently contain reference sections, maps, glossaries, computer printouts, photographs, extended descriptions of methods, and lengthy comparative data. Appendixes are both self-contained and closely connected to the body of the report.

If the appended information is of more than one kind, create two or more appendixes, each identified by letter and title, for example:

Appendix A. Glossary of Terms
Appendix B. Site Maps
Appendix C. Project Cash Flow
Appendix D. Equipment Specifications

Number the pages within each appendix with the appropriate prefatory letter (page B-3, for example, will be the third site map of Appendix B). Number figures, tables, and equations with the letter designation of the appendix in which they appear. Start each appendix on a new page (Figure 10.4).

A good set of appendixes is not a dumping ground for leftover figures and calculations but the location for particularly complex and specialized data that will be crucial for some readers—probably the most important readers. Make explicit connections between appendixes and report body. For example, when you discuss a proposed wind energy farm, *tell* readers that site maps are located in Appendix A and equipment specifications in Appendix B. List the title of each appendix in the table of contents and consider preparing a second table of contents to material in the appendixes, located just before Appendix A.

References

(1) Clark, L., and Baxter, K. M. Groundwater sampling techniques for organic micropollutants: UK experience. *Q.J. Eng. Geol.*, 1989, **22**, 159-168

(2) Department of the Environment, The use of herbicides in non-agricultural situations in England and Wales, *Foundation for water research*, 1991

(3) Klint, M., Arvin, E., Jensen, B. K., and Snijders, A. Biodegradation of the pesticides atrazine and MCPP aquifers, Tech Univ. of Denmark, 1990

(4) Lawrence, A. R. and Foster S. S. D. The pollution threat from agricultural pesticides and industrial solvents, *BGS Hydrogeological Report 87/2*, 1987

(5) Pionke, H.B., Glotfelty, D.E., Lucas, A.D., Pesticide contamination of groundwaters in the Mahantango Creek watershed. *J. Environ. Quality*, 1988, (1) 76-84

A-1

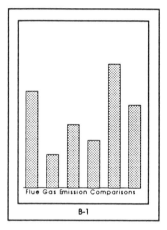

B-1

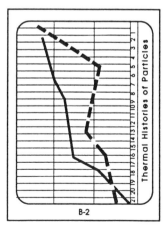

B-2

Figure 10.4

Appendix A in this document contains References. Appendix B contains two figures. Note that the figure on page B-2 is in landscape orientation.

Audiences for Reports

Readers of formal reports nearly always represent a complex, varied audience with different purposes, different amounts of time to spend on the document, and different information needs. A cost accountant reading a report that recommends the substitution of geothermal steam for conventional electricity is more interested in the cost sections than in the technical sections that specify details about deep-drilling equipment. An environmental impact analyst may consult only the executive summary, the environmental impact section, and selected appendixes that amplify information about affected wildlife. Very few readers will need to read every section of a long report. In general, early sections of reports are less technical than later sections, and appendixes are usually directed at specialists.

Reports need to be constructed so that varied audiences, with varied purposes for reading, can chart their own reading paths. Folk wisdom says that 80% of readers will read only 20% of any document. You cannot therefore shortchange any part of the document. Instead, you need to make each section strong and self-sufficient because it may be the only section an important reader reads.

Methods in Academic Laboratory Reports
Reports prepared in research and development settings usually focus on results, but in academic laboratory settings, researchers and students need to write about how they *got* the results. A laboratory report usually provides more space for an account of how you did it than of what it means, though a good laboratory report will always draw conclusions and suggest interpretations.

Typical elements in student laboratory reports are displayed in Figure 10.5 and can serve as guidelines for format. Like any other technical document of more than a page or two, a laboratory report will be improved by the addition of a cover page, a table of contents, a list of illustrations, and one or more appendixes to present important but unusually detailed or specialized data like computer code or extended calculations.

LABORATORY REPORT FORMAT

Front Matter

Title Page
Abstract
Table of Contents
List of Figures and Tables
List of Abbreviations and Symbols

Report Body

Introduction
Theoretical Background
Equipment List
Experimental Procedure
Results
Analysis
Conclusion

Appendixes

References
Supporting Details

Figure 10.5
Conventional components of laboratory reports.

Decision-Making Processes in Recommendation Reports

Recommendation reports evaluate a process or a product and recommend (or perhaps advise against) a specific course of action. Typically, they begin with an explicit statement of what they recommend. They then specify the criteria that have served as the bases for judgment, and they compare alternatives against criteria. Finally, they list the action required to implement the recommendation (Figure 10.6). Recommendation reports are the stock-in-trade of consulting engineers, who use their special expertise to make informed recommendations to others.

Because recommendation reports serve as the bases for decisions, they must provide explicit information about the criteria that have informed the author's thinking. What are they? Why do they matter? In a report

RECOMMENDATION REPORT FORMAT

Front Matter

Letter of Transmittal
Title Page
Executive Summary
Table of Contents
List of Figures and Tables
List of Abbreviations and Symbols

Report Body

Recommendation
Introduction
Decision Criteria
Analysis of Alternatives
Conclusion
Action Required

Appendixes

References
Supporting Details

Figure 10.6
Conventional components of recommendation reports.

recommending the location for a new manufacturing plant, for example, criteria for comparing potential locations might include proximity to railroad tracks, state and local tax rates, and available skilled labor pool. A comparison of competing alternatives against the same criteria results in a tradeoff analysis. Candidate Plant Location A may be conveniently located with respect to railroad tracks and an outstanding pool of available skilled labor. The tax rate for A, however, may be so prohibitive that Candidate Plant Location B, with its low tax rate, is preferable despite its great distance from railroad tracks and a marginal pool of skilled labor.

For many problems, one course of action will not be clearly superior to another. In such cases, do not attempt to oversimplify; a frank discussion of your concerns will be more welcome than an attempt to conceal discrepancies. In some research settings, authors of recommendation reports are *required* to include a subjective rating of their confidence in the chosen alternative.

Executive Summaries for Management

Because managers and other decision makers are unlikely to be interested in the detailed technical analyses of a problem, management reports begin (rather than end) with summaries, conclusions, and recommendations. The executive summary is pitched at readers who may lack the technical expertise to follow particulars of the work but are interested in the implications of the report. In the foreground is concern with what the results mean and what actions should now be taken.

Executive summaries, often 5 to 10 pages in length, contain the ideas of the report in semitechnical terms, and they replicate the format of the report in miniature, with headings and illustrations. Sometimes they are detached from the body of the report and bound separately for distribution to a much wider audience than the report will have (Figure 10.7).

The tone of an executive summary is typically persuasive, sometimes even enthusiastic, with emphasis on the importance of the problems addressed. Executive summaries stress benefits, conclusions, and recommendations rather than the way the work was done. Their tone approximates the sales rhetoric of proposals, rather than the cautious optimism of progress reports. An executive summary should be self-contained and complete, so that readers are not directed to pages of the report but have all the information they need to understand the main findings.

Managing Complex Report Writing and Production

Planning for Coauthorship and Deadlines

Managing the group writing process for a final report is much like the process of proposal management. Report writing will present the same task allocation and time management problems. The importance of a well-thought-out, section-by-section document plan should be obvious to any research group that has got this far.

EXECUTIVE SUMMARY

Value Engineering with
Existing Infrastructure:
An Example Using
Wind Energy on the
Antioch Bridge

Mel Manalis, Ph.D.
Jim Davidson

Caltrans Contract 53Q386
December 22, 1992

Introduction

The two mile-long Antioch Bridge spans the San Joaquin River as it flows westward towards the Carquinez Straits in a wind-swept environment. The bridge structure amplifies the wind speed as the air flows around the bridge. We were intrigued with the idea of exploiting this amplification of wind speed to produce electricity from wind turbines mounted on the bridge. Would bridge-mounted wind turbines be able to produce electricity from this flowing air at an unusually low and yet compatible with bridge structural and environmental requirements?

This study identifies the technical, economical, and environmental issues germane to answering the above question. We directed considerable effort towards determining the spatial and temporal variations of the wind speed surrounding Antioch Bridge since the energy content of flowing air is proportional to the air speed cubed. We began anemometer studies on June, 1988 to search for and document areas suitable for electricity production from wind turbines affixed to the bridge. Subsequently, about 180,000 hours of wind energy data from several anemometers located on and near the bridge have been recorded and analyzed.

1

Discussion

1.   Wind turbines are to be located at areas of maximum wind energy underneath the bridge, away from auto, maritime, and foot traffic.

2. The amount of wind-generated electricity that can be produced annually from all the amplification zones under the entire bridge is 26 times the electricity consumed by the toll plaza headquarters.

3. Eighty percent of the electricity is generated from the bridge-wind turbines in the summer, when the utility's demand for electricity is the strongest.

4. Wind turbine characteristics are selected to maximize compatibility between bridge structure technology and wind technology.

5. Revenues, costs, and investment incentives associated with wind-generated electricity at the bridge indicate potential for considerable cost savings.

2

Recommendations

1. Interact with Caltrans' Office of Structure and Design to explore the bridge-turbine interfacing issues delineated above.

2. Develop new contracting possibilities with representatives of PG&E regarding offsetting electricity consumption at the toll plaza headquarters with electricity generated from wind turbines located under the bridge. Perhaps these discussions could address future electricity demand for lighting the bridge.

3. Examine the compatibility of bridge-supported wind turbines with the Sherman Island Wildlife Management Plan.

3

Figure 10.7
The authors of this executive summary include an overview of their project, a brief discussion of findings, and three recommendations. They prepared the executive summary as a separate bound volume, and they included relevant maps and illustrations (not shown here). (Courtesy of M. S. Manalis and J. Davidson.)

There *will* be a deadline, a time by which the report must be finished and delivered. Schedules, calendars, shared understandings, and frequent group meetings are just as important here as in the completion of any other complex task. The contributions of group members need to be coordinated.

Groups should meet regularly through the research work and should consider the writing required as well as the progress of the technical work at every meeting. When the time comes to write the final report, no group member should have a clean slate, needing to start from scratch to construct sections of the document. Although no report will write itself from assorted laboratory notes and other records kept by project investigators, a report can be more efficiently and effectively written if investigators do not separate the writing tasks from the other investigative tasks of the project.

Distributing Writing and Format Guides

Many active research and development settings have an official style guide available for everyone to use (Figure 10.8). By consulting the style guide, engineers, scientists, writers, editors, technical illustrators, and others involved in document production do not need to create and learn a new set of rules for each project. If your report follows from a group of earlier documents, you will want to plan carefully, so that the style guide for the final report has the same features as earlier documents on the subject. Standard publications practices may take time to develop initially, but the long-term payoff is saved time and improved consistency of style and appearance.

The most essential task in document design is to make sections and pages predictable. Even a simple document template will provide the same kind of information, in the same place, in the same type style on every page. (See chapter 16 for guidelines on page and document design.)

Editing for Clarity and Accessibility

Some sections of reports have a life of their own. As you are writing, remember that titles, keywords, and abstracts may be indexed in technical databases and become part of the literature of science and engineer-

```
                    NUCLEAR DIVISION
              DOCUMENT PREPARATION GUIDE          CHAPTER   6
                  Contract W-7405-eng-26          PAGE      6-65
     ─────────────────────────────────────────────────────────
     SUBJECT:  PREFERRED USAGE
     ─────────────────────────────────────────────────────────

        hectare ...................................................... ha
        henry ......................................................... H
        hertz (singular and plural) ........................... Hz
        hexagonal close-packed ................................ hcp
        high voltage ............................................... hv
        horsepower ................................................. hp
        horsepower hour ......................................... hp·h
        hour .......................................................... h
        hundredweight ............................................ cwt
        hyperfine structure .................................... hfs
```

Figure 10.8
Writers in the nuclear division of one government installation refer to the *Document Preparation Guide* distributed by the publications manager. The guide covers a wide range of issues, including abbreviations and punctuation.

ing, independent of the report document. Executive summaries will not be indexed, but they often have a much wider distribution than the document itself. Sometimes they are printed and distributed separately, the only section that important readers will read. Needless to say, all report sections need attention and care. But sections that will stand alone should be the focus of particularly thoughtful preparation and vigorous editing: they must pass the tests of clarity and self-sufficiency.

Many report writers deliberately invest the largest amount of editing time on the most widely read report features: titles, tables of contents, abstracts, executive summaries, headings, and legends to illustrations. Consider nonverbal as well as verbal features when you edit. You may want to add tabbed dividers to increase the ease with which readers can find exactly the information they need, or you may add color to highlight key points in complex illustrations. Consistent use of white space to create separations between major ideas will increase the readability of your report. White space is a particularly important feature if your report is scanned or republished in electronic format.

Bringing all Drafts and Boilerplate Up-to-Date

By the time you write a final report, you probably have both text and illustrations that you can update and reuse. If you have labeled and maintained data files carefully, you may save yourself needless duplication of effort. But allow time to review previously written material (often called boilerplate) with great care: a mechanical cut-and-paste effort will not be enough to assure currency and consistency. Figures and tables will always need to be renumbered from document to document, and text references to figures and tables will need to be updated accordingly. Heading styles may need alteration for consistency. Verb tenses usually need attentive editing from proposal to final report.

Using Headings to Map Report Structure

Because headings stand out from the text, they send powerful signals about the relative importance of the material that follows. Their size, placement, and typographical features, such as bold or italic type, are therefore important. Establish a style for headings and always take the time to edit a report for consistency in heading style.

Even generic section and subsection headings like Background, Process Description, and Conclusion will help nonspecialist readers recognize parts of a report. But reports aimed primarily at nonspecialist readers can be even more effective if headings and subheadings are informative. For example, "System Efficiency and Maintenance" is more specific than "Technical Criteria." When the exact wording of headings is presented in the table of contents, readers have an easily accessed, helpfully detailed overview of the report.

Making Graphics Accessible

Tables and figures—collectively called illustrations in scientific and engineering writing—take up a large percentage of the pages in most reports. In fact, many reports contain more pages of tables and figures than of text. Every illustration is a piece of technical literature in its own right, one that may be more widely studied than the text of the report and that may actually circulate separately and be reused in another report.

Every table and figure should have a title or legend: legends (usually called titles in tables) are set *above* tables and *below* figures. Legends should be informative and clear—many report readers skim the text and read illustrations and legends with extreme care. Legend style should be consistent throughout a report: either full sentences or sentence fragments.

Every table and figure should be numbered. For long reports, illustrations are usually numbered according to the section of the text in which they occur (Figure 4.2, for example, is the second figure in the fourth section). In shorter reports, figures and tables are usually numbered straight through. All tables and figures that are derived from sources must mention their source in the table footnote and figure legend, or at the bottom of the illustration, set close enough so that the illustration never gets reproduced without the reference.

All tables and figures should be referred to with the word *Table* or *Figure* and the designated number. Some writers like to integrate the reference into the report text; others prefer to use parentheses:

The design takes advantage of a commercially available array of solar collectors (see Figure 6.12).

or The design, illustrated in Figure 6.12, takes advantage of a commercially available array of solar collectors.

Studies of the ways that technical readers read reports show that parentheses seem to serve an important function, guiding readers *back* to the place they left in the text when they moved ahead to look at an illustration.

Placement of illustrations presents extreme challenges. The conventional advice is that illustrations should be located directly after the first text reference, but this directive is often not very useful. By the time you have placed Table 2.3 or Figure 3.4, you usually find that text and illustrations are no longer on the same page. If you are committed to text and illustrations being closely connected, you can use storyboard format, consistently presenting text on the left-hand pages and illustrations on the right, or you can prepare a separate volume of illustrations so that readers can read text and illustrations side by side (see chapter 16 for alternative ways to present large-size tables and figures in bound reports).

Erratum Sheet

On page 19:

Table 6, Ocean Disposal of Low-Level Radioactive Wastes, 1957-1969

should read:

Table 6, Ocean Disposal of Low-Level Radioactive Wastes, 1967-1969

Figure 10.9
This erratum notice alerts readers to an error in the report text and provides corrected information.

Informing Readers about Errors

If you discover serious errors in your text or illustrations after the report has been printed, prepare an erratum sheet (for one error) or an errata sheet (for more than one). List the page number and any other information that will help locate the error, such as the line number or figure number (Figure 10.9), and slip the sheet into the report document, between the front cover and the title page.

Publications beyond the Report

A completion report is rarely the first document on a subject, nor is it always the last. In the course of work in science and engineering, research results may be repackaged and disseminated in oral and written forms: at meetings and conference presentations; in conference proceedings, reports, and refereed journal articles. Chances are that portions of your report will appear in other documents, and your results will form the basis for further research.

11

Journal Articles

An academic research team spends 3 years on a project and, with some changes in direction, reports results that please both the team members and the funding agency. With the work complete, most of the team plans to move on to other projects. The team leader, however, has one more project task in mind: final results should be published in a reputable scientific journal. Publication will take little effort, the leader assures the rest of the team. All that remains is to rewrite the final report so that it meets the specifications of a journal article. Promotions, the team leader points out, may depend on this venture.

Advancement in science and engineering is frequently tied to publication of research in refereed journals, where articles submitted for publication are reviewed by several experts ("referees") who assess the validity and originality of the work. With journal publication, your contribution to a field is no longer informal. Your work now lies in the formal domain of technical literature. Once published, your work is accessible through secondary sources and becomes part of the knowledge in a field. After publication, your article may be cited by others working in the field. Colleagues may evaluate its contribution and link new information to it. Your published journal article can become a reference from which new theory is advanced and new evidence added.

The first release of information on any technical subject is unlikely to appear in a refereed article. The delay from the start of a project to journal publication may be 3 years or more, so that by the time a journal article sees print, the authors have probably discussed the research in a variety of written and oral forms. In written form, the information may have appeared in one or more proposals, progress reports, completion reports, papers for conference proceedings, and in some fields (physics, for one), a brief form of journal publication called a "letter." The same information will probably have been discussed informally and may have been the subject of presentations at professional meetings. This sequence of activities helps authors to shape their work for journal publication.

Furthermore, once an article manuscript has been completed, it is considered a preprint and may be circulated widely to interested researchers in the "invisible college" of people working on similar projects that makes up most fields. Some authors do not distribute preprints until the article has been accepted for publication; others distribute preprints to obtain feedback and suggestions for modifying the manuscript.

The practice of widely distributing preprints is at least partly a response to the great amount of time required for a journal article to appear in print. Some scientists and engineers claim to read preprints only, because journals do not contain current views on specific research problems. Preprints are often circulated by way of e-mail, increasing the speed with which they reach interested readers.

Targeting a Journal for Submission

Estimates vary, but there may be 50,000 to 70,000 refereed science and engineering journals published at the present time. Still, the nature of your research and the scope of your paper will limit rather dramatically the number of journals to which you might submit your work. Journal titles indicate the general area of interest, and many journals also include more explicit statements of their editorial policy (Figure 11.1). Your chances of acceptance are enhanced if your subject is closely matched to the research interests of the journal. Do not send manuscripts to a journal without first carefully considering whether the scope of your work fits the journal's publishing profile. Always analyze the journal in which you hope to publish.

Though some journals publish only one kind of manuscript, many publish contributions in more than one category. Types of manuscripts may include:

Instructions to Contributors

AIChE Journal will consider unpublished manuscripts dealing with significant theoretical and experimental developments in chemical engineering. Four types of manuscripts are printed: Letters to the Editor; R&D Notes; Papers; and Review Papers. Letters to the Editor may concern previous *Journal* articles, comment on research trends, or offer constructive suggestions toward improving the *Journal*. R&D Notes are limited to 8 double-spaced typewritten pages including figures and tables. They receive one review. Papers are limited to 35 double-spaced pages including figures and tables. They receive three reviews. Review Papers cover extensively an important area in chemical engineering. Authors planning a review paper should consult the editor for requirements.

Materials Science and Enginering provides an international medium for the publication of theoretical and experimental studies and reviews of the properties and behavior of a wide range of materials, related both to their structure and to their engineering application. The varied topics viewed as appropriate for publication include but are not limited to the properties and structure of crystalline and non-crystalline metals and ceramics, polymers, and composite materials. Types of contributions include original research work not already published, plenary lectures and/or individual papers given at conferences, reviews of specialized topics within the scope of the journal, engineering studies, and letters to the editor.

Figure 11.1
Most journals provide instructions to contributors, with explicit information about research subjects that are suitable for publication.

• *Articles or research papers.*   These reports of original research work are usually assessed by at least two independent referees.

• *Letters.*   Brief communications are sometimes judged to be both timely and important. These are usually no more than 2500 words in length, and in the interests of rapid publication, many journals send letter manuscripts to only one referee.

• *Notes.*   Journal notes elaborate on previous papers published in the journal, present new experimental data, or develop a new theoretical concept. These manuscripts often receive only one review.

• *Reviews.*   Critical reports survey recent developments in a field and are usually commissioned by the editor.

• *Scientific correspondence and letters to the editor.*   Space is usually reserved for discussion of papers previously published in the journal and for miscellaneous topical issues.

Conventions of Refereed Articles

Writing a high-quality paper for publication in a refereed journal takes more than following a formula or filling in the blanks. The refereed journal article develops as a stylized assemblage of sections, each devoted to a specific kind of content: theoretical, methodological, empirical, and interpretive (Figure 11.2).

The late Nobel biologist/physician Sir Peter Medawar argued eloquently in 1964 that the scientific paper is a fraud. Medawar claimed that its inductive structure fails to represent the processes of thought that lead to scientific discoveries. In Medawar's view, the discussion (which traditionally comes at the end) should come at the beginning, followed by the theory and methods. Nevertheless, many researchers claim that the act of writing a journal article is a significant part of the science. During the process of writing a manuscript, authors often discover gaps in experiments. The writing then sends them back to the laboratory for more experiment and analysis. Recognize that in the writing of even the simpler elements of your article you may find that you have further work to do.

Front Matter

Title   The title will be more widely read than any other part of the article. Titles allow potential readers to judge the relevance of the document

JOURNAL ARTICLE FORMAT

Front Matter

Title
Abstract
Keywords

Article Body

Introduction
Theoretical Section
Experimental Section
Results
Discussion
Conclusion

End Matter

Acknowledgments
Reference
Appendixes

Figure 11.2
Standard elements in a journal article.

for their own interests. Titles also provide indexers with keywords to use as they prepare subject indexes for bibliographic reference services. Your title should be concise and informative, reflecting the specific content of your work, emphasizing keywords, eliminating filler words. For example, rather than "Survey and Analysis of HIV-Induced Immunodeficiency Caused by Programmed Cell Death of Reactive T Cells," your title might read "Programmed Death of T Cells in HIV-1 Infection."

The editors of the journal *Nature* provide the following instructions for titles: "Titles say what the paper is about with the minimum of technical terminology and in fewer than 80 characters. Active verbs, numerical values, abbreviations, and punctuation are to be avoided. Titles should contain one or two keywords for indexing purposes."

Abstract   After the title, the abstract is more widely read than any other part of an article. The abstract is a stand-alone piece of technical literature, not an introduction to the article but a capsule version of it, the article in miniature. The vocabulary of the abstract is likely to serve as the basis for bibliographic searches on your subject. An abstract that clearly and accurately represents the problem you addressed, your methodology, and main results will assure that future researchers in your area learn of your work.

Write the abstract for your article in accord with instructions (or examples) provided in the journal to which you are submitting the manuscript. Abstracts vary in their length and content, but they are typically 150 to 200 words. Some—called descriptive abstracts—give only a general idea of what the article covers. Others—called informative abstracts—include greater detail (Figure 11.3).

Keywords   For other researchers in your field, the keywords that identify your subject and focus will be crucial access routes to your publication. The journal in which you publish may ask you to select keywords from a predetermined list, or you may be free to select your own terms. As Figure 11.3 shows, you will naturally want to select words that represent the most important concepts in your article.

Article Body

Introduction   The main function of the introduction is to identify the objectives and rationale for your paper. The introduction argues for the originality, the good antecedents, and current connections of your work. This first section establishes that you are, as historian of science Derek deSolla Price so trenchantly put it, "running with the pack and trying with them to do something new."

The introduction specifies the problem addressed, briefly summarizes previous research, and indicates what the present work will add to what is known about the subject. The last paragraph of an introduction typically contains one or two sentences that serve as both preview and summary of the subject and findings of the paper.

Energy Sources, Volume 13, 121-135

Enhanced Recovery of Bitumen by Steam
with Chemical Additives

V.A. ADEWUSI
Dept. of Chemical Engineering
Obafemi Awolowo University
Ile-Ife, Nigeria

ABSTRACT *Enhanced recovery of bitumen by a combination of steam and caustic chemicals with and without alcohol has been investigated using both static and dynamic techniques. Static experiments were conducted on the basis of information derived from a previous investigation to define the alcoholic caustic formulations potentially effective for bitumen mobilization. The aliphatic alcohol that gave increased bitumen emulsification with a further indication of a more favorable condition for bitumen microemulsion displacement was then selected for the dynamic tests, in which a caustic concentration level up to 0.4 wt% was propagated by a saturated steam at 400 kPa through oil-bearing sand packs. Data obtained from these tests revealed a better bitumen displacement and greater sweep efficiency in steamfloods with caustic and 50% octanolic-caustic additives than in steam drive. In addition, bitumen production began earliest in the alcohol system and was accompanied by increased production rate, leading to more than 10 and 30% increases in the overall bitumen recovery compared with the nonalcoholic alkaline steam and pure steam injection processes, respectively. These results also served to reinforce the findings of the static experiments.*

Keywords bitumen microemulsion, bitumen recovery, steam drive, steam-caustic process, steam-caustic-octanolic process

Figure 11.3
Here an informative abstract summarizes objectives, methods, results, and discussion. It serves as the article in miniature. In contrast, a descriptive abstract gives only a brief and very general idea of the subject. (Source: Adewusi 1991.)

Theoretical Section   Sometimes you can provide enough theory in the introduction to support the paper and move directly to the experimental section. But in many papers, more extensive discussion is necessary, and you will have to provide a separate theory section. The theory section may contain a predictive model or series of governing equations (though detailed mathematics should go in an appendix). It may contain a survey of design parameters, a discussion of assumptions, or a historical survey of previous work. These discussions frame the topics and variables that will be the main subjects of your experimental, results, and discussion sections.

Many articles and reports do not develop their own models but rely on other published work. In these cases, the authors cite the papers that developed the original models. Thereafter, they draw on equations from the cited papers and assume that readers interested in a full development of the models will refer to the original works.

Experimental Section   The experimental section of an article describes the tools and processes that enabled you to meet the stated objectives of the introduction. This section is sometimes called materials and methods, experimental methods, procedure, or experimental apparatus, depending on the stylistic preferences of the journal. The section will be read for at least two major reasons. First, readers will judge how skillfully you have designed the empirical processes of problem solving. Second, readers may test your methodology against your results in their own laboratories. In experimental sections, clarity and accuracy are priorities. You are describing a variety of objects and processes that have been used to deliver a set of data. Include significant numbers, but move detailed analyses to appendixes.

Results   The results section translates the empirical terms of the laboratory into the language of numerical generalization and statistical analysis. All sections of a journal article lead up to or away from the results section, and the results section may retain its value long after the methods and conclusions have become obsolete. Results are confined to their own section not only because they manifest a distinct phase of your research but also because readers often like to work their own interpretations on data, perhaps considering alternative conclusions. When results are mixed with interpretations, the integrity of the data can be compromised.

Data may be presented in several ways. If results are simple, you may be able to note them in a brief prose passage. Series of data should be presented in graphic form. You do not need to be exhaustive: condense data according to standard analytical procedures into meaningful representations of your work. Avoid merely noting that the data are shown in a table or figure: draw out the importance of trends shown in each illustration.

Discussion In the discussion section, you evaluate your results and their significance. Just presenting results is not enough. Data rarely speak for themselves. The discussion section may note discrepancies in the findings and explicitly discuss the reliability of the results. By identifying inconsistencies and noting their sources, you lend credibility to your work. Even when you have no clear explanation for inconsistencies, you should note their existence.

Conclusion In the conclusion, you can restate your findings and assess their implications. In this section you may specify possible applications of your findings and, if appropriate, recommend directions that future research on the topic should take. These statements should bring you full circle to the original problems and objective of the work. They identify your main accomplishments and connect your work with larger issues.

End Matter

Acknowledgments Most journal articles end with a brief section of acknowledgments. It is good practice to ask individuals for permission before including their names, though it is expected that you will acknowledge any funding agency that supported your work.

References Several reference styles are used in scientific and engineering publication (see chapter 18). Base your style on the guidelines provided in the journal. If these are not explicitly stated, you can derive stylistic guidelines by examining published papers.

ENERGY SOURCES
Instructions to Authors

Manuscript. Manuscripts will be accepted with the understanding that their content is unpublished and not being submitted for publication elsewhere. All parts of the manuscript, including the title page, abstracts, tables, and legends should be typewritten, double-spaced on one side of white bond in English. Allow margins of at least one inch (3 cm.) on all sides of the typed pages. Number manuscript papers consecutively throughout the paper. Translations of papers previously published in languages other than English can be considered, but this information must be provided by the author at the time of submission.

Title. All titles must be as brief as possible, 6 to 12 words. Authors should also supply a shortened version of the title suitable for the running head, not exceeding 50 character spaces.

Affiliation. Include full names of authors, academic and/or other professional affiliations, and the complete address of the author to whom proofs and correspondence should be sent on the title page.

Abstract. Each article should be summarized in an abstract of not more than 150 words. Avoid abbreviations, diagrams, and reference to the text.

Keywords. Authors must supply three to ten key words or phrases that identify the most important subjects covered by the paper.

References. Citations in the text are by author(s) and date in parentheses. Full citations should be arranged alphabetically and must conform to *The Chicago Manual of Style, 14th ed.*

Journal article:
Gunn, E.S., S.C. Ballard and M.D. Devine. 1984. The Public Utility Regulatory Policies Act: Issues in Federal and State Implementation. *Policy Studies Journal* 13(2):353-364.

Book:
Keeney, R.W. 1980. *Siting energy facilities.* New York: Academic Press.

Figures. All figures must be submitted in a camera-ready form. Xerox copies are not acceptable. Figures must be submitted either as black and white glossy photographs or photostats (bromides). Label each figure with article title, author's name, and figure number by attaching a separate sheet of white paper to the back of each figure. Do not write on the camera-ready art. Each figure should be provided with a brief descriptive legend. All legends should be typed on a separate page at the end of the manuscript. Within the text spell out "Figure" on first reference.

Tables. All tables must be discussed or mentioned in the text and numbered in order mentioned. Each table should have a brief descriptive title. Do not include explanatory material in the title: use footnotes, keyed to the table with superior lower-case letters. Place all footnotes to a table at the end of the table. Define all data in the column heads. Every table should be fully understandable even without reference to the text. Type all tables on separate sheets; do not include them within the text.

Permissions to Reprint. If any figure, table, or more than a few lines of text from previously published material are included in a manuscript the author must obtain written permission for republication from copyright holder and to forward a copy to the publisher.

Page proofs. All proofs must be corrected and returned to the publisher within 48 hours of receipt. If the manuscript is not returned within the allotted time, the editor will proofread the article, and it will be printed per his instruction. Only correction of typographical errors is permitted. The author will be charged for additional alterations to text at the proof stage.

Offprints. A total of fifty (50) offprints of each paper will be supplied free of charge to the first named author unless otherwise specified. Additional copies may be ordered at charges shown on the offprint price scale, which will be sent to the author with the proofs. Offprints are not available after publication of the issue.

Appendixes Though journal articles are usually brief accounts of research findings, they sometimes contain appendix sections. In the *AIChE Journal* (American Institute of Chemical Engineering), an appended section of notations regularly follows the acknowledgment section for each paper. In other journals, detailed accounts of methods or unusually specialized results are presented as appendixes to the paper.

Manuscript Preparation

Your chances of acceptance are enhanced if your manuscript is prepared in the style preferred by the journal. Most editors provide detailed instructions for manuscript preparation, often inside the back cover of each issue under the heading "Instructions to Authors." Here you will find guidance about manuscript style, equation style, number of copies required, reference style, and instructions for preparation of tables and figures. In all cases, follow journal guidelines. Most journals ask that you submit copy for illustrations as well as lists of figure legends and table titles on separate pages. Some specify size limits for tables and figures; some announce page charges, with extra page charges for color artwork (Figure 11.4).

Some publishers are now supplying authors with electronic instructions and templates. The American Mathematical Society, for example, provides author packages for 27 of its publications. Each package includes instructions, author handbook, style files, templates, and samples (http://www.ams.org/tex/author-info.html).

Submission and Resubmission

Except for commissioned review articles, editors will rarely invite you to publish your research (though this happens). More likely you will submit your manuscript, accompanied by a one-page letter of transmittal

Figure 11.4
Most journals provide manuscript preparation instructions to authors. Instructions for preparing artwork and references are particularly important, and they vary among journals. (Source: Energy Sources, New York: Crane Russak and Co.)

addressed to the journal editor (Figure 11.5). In this letter you should state that you want to have the paper—identified by its title—reviewed for publication. In a second paragraph you should present the main focus of the article. Conclude the letter by thanking the editor for considering your request. If a copyright waiver form is required for publication, it should be enclosed with your letter; many journals provide such a form in each issue. Some journals encourage authors to suggest names of several experts in the field who would (or who would not) be good referees, and you can include such information in the letter of transmittal. Some journals ask that you include a statement confirming that the manuscript has not been published previously and is not also being considered for publication in another journal.

The editor will screen the paper to ensure that its subject is appropriate for the journal. At this stage, your letter of inquiry may elicit a simple rejection letter in response: your paper has not even been reviewed because its subject matter falls outside the range of the targeted journal. More likely, you will have considered this issue before you shipped off your paper, and your manuscript will be sent out to several expert, anonymous reviewers who will independently assess the adequacy of the science, the significance and originality of the project, and the clarity of presentation. Reviewers then return their comments to the editor (Figure 11.6). Sometimes an author must resubmit a manuscript several times before achieving publication, but critical feedback from referees often improves the manuscript.

The editor will respond in one of three ways:

• *Rejection.*   In a letter of rejection the editor may include excerpted comments from reviewers or a summary of those comments. The comments may be of help in revising your paper for submission to another journal.

• *Conditional acceptance.*   On the basis of reviewers' comments, the journal is interested in publishing your article but only if certain changes are made. The editor delineates those aspects of the paper that, in the estimation of the reviewers, need revision. This type of acceptance is the most typical. Few articles are accepted outright.

• *Acceptance.*   On the basis of reviewers' comments, the journal publishes your paper as submitted—a rare occurrence.

December 15, 1999

A.R. Nathanson
Editor, *Tidal Energy*
Dept. of Ocean Engineering
University of California
Santa Rosita, CA 93106

Dear Professor Nathanson:

Enclosed are three copies of an original paper, "Design Tradeoffs for
Tidal Mini-Hydro Power Plants," submitted for review in **Tidal Energy.**

The study compares three basic methods of operating a tidal
hydropower system: tide-cycle, ebb-cycle, and double-cycle. The
manuscript has been prepared in accord with your Instructions to
Contributors. The material has never been published, and it is not
under consideration for publication elsewhere.

The co-authors are Brian Kato, Barney Greinke, and Peter Campbell of
CGK Engineering. The research reported in the paper is part of an ongoing study of the feasibility of tidal mini-hydro power plants.

Thank you for your attention to this paper. We look forward to
receiving your comments.

Yours truly,

Brian Kato

Brian Kato

BK:sw

Enclosure

2535 West Armadillo • Suite 205 • Santa Rosita • California • 93106

Figure 11.5
This letter of transmittal accompanies three copies of a manuscript submitted for
journal publication. The letter explicitly states that authors have read and followed the instructions to contributors.

Studies in Earth Science
CONFIDENTIAL REVIEWER'S REPORT

1. Is the subject of the paper consistent with the scope of the journal?	YES	NO
2. As far as you know, has this material been published before in English?	YES	NO
3. Does the scientific content justify the space it will occupy?	YES	NO
4. Can any parts of the paper be shortened or omitted without loss of scientific content?	YES	NO
5. Are any errors of fact or logic contained?	YES	NO
6. Are all figures necessary?	YES	NO
7. If the paper contains graphs and tables based on the same data, is it necessary to include both?	YES	NO
8. Does the abstract (normally 50-150 words) bring out the main points of the paper?	YES	NO
9. Is the title suitable and adequate?	YES	NO
10. Are the literature references adequate?	YES	NO
11. What is your overall recommendation?	YES	NO

____ PUBLISH AS SUBMITTED ____ DO NOT PUBLISH

____ PUBLISH WITH MAJOR REVISION ____ PUBLISH WITH MINOR REVISION

COMMENTS: Please use a separate sheet to expand on the above and to suggest changes that you feel would improve the manuscript, in particluar with regard to its length.

Signature: ____________________ Date: ____________________

Figure 11.6
Reviewers of journal articles typically return a checklist and a separate sheet of detailed comments to the editor. The journal mentioned here is invented, but the form of the confidential reviewer's report is typical.

If you receive a letter of conditional acceptance, you may decide to follow the editor's comments and change certain parts of the paper, or you may decide to withdraw the paper from further consideration. In either case, you must respond in a timely fashion with a letter stating your decision. If you decide to change your paper, your next submission will be reviewed again, either by the same referees or by the editor alone. You may need to go through several cycles of submission and review before the editor considers your paper ready for publication.

The former editor of *Science*, Philip H. Abelson, takes an optimistic view about the chances of a paper being published. Although *Science* accepts only about 20% of manuscripts submitted, Abelson believes that almost all of the rejected papers appear eventually in other journals. The editor-in-chief of the American Physical Society (APS), David Lazarus, reports an approximate 60% acceptance rate for the APS journals: 15,000 manuscripts submitted annually, with approximately 9000 accepted. Recognize, of course, a difference in prestige among journals. An article rejected by prestigious research journals but finally published in an unrefereed newsletter will have substantially less influence on the practice of science and the progress of a career than if it had been published in the originally selected research journal.

After the paper has been accepted, it may be copyedited for stylistic consistency and correctness by an editorial assistant on the journal staff. Some journals will, at this time, ask you to submit corrected copy ready for publication (called "camera-ready" copy). If the journal takes responsibility for typesetting, your next task will be proofreading. Journals that levy page charges for publication may ask you to submit them when you return the proofs. You can also, at this time, order extra offprints.

Collaboration on Journal Articles

Research in science and engineering is rarely a solitary task; most published papers have two or more authors. Coauthors need to ensure that each person listed participates fairly in planning, writing, and revising; developing and monitoring a work schedule; and editing to eliminate differences in style.

Coauthors may work in the same laboratory or university, but they are frequently located at great distances from one another and make substantial use of telephones, e-mail, and fax machines. Successful coauthoring usually requires at least an initial face-to-face meeting to brainstorm, plan, and divide the work, with explicit discussion of ways to coordinate activity and monitor progress.

The coauthors may decide to divide the work by giving each researcher primary responsibility for one or more sections of the paper, designating one person with responsibility for combining and editing. Or they may designate one person to write a draft of the entire paper, while all other members of the group function as editors. Or they may give responsibility for tables and figures to one or more group members and responsibility for text to others. Each of these methods works for some groups and not for others. The critical factor in successful collaboration is shared understanding of the purpose and content of the document so that each writer knows the larger context into which a contribution fits.

Computer conferencing may take the place of face-to-face meetings of collaborators at remote sites. A simultaneous teleconferencing facility enables several authors to view the same display on their respective workstations while they work on the same underlying data structure. As a draft of the manuscript is assembled, reviewing editors can make on-line annotations by creating a node with comments linked directly to the section being discussed.

Electronic Journals

The refereed journal system is in transition, partly because electronic access is creating new possibilities and partly because printed journals do not provide adequate speed, access, or economy. Delays from submission of papers to publication mean that refereed, archival journals cannot help researchers keep up with new developments. To anyone doing serious work in an active scientific or engineering field, information in journal articles is relatively old. The volume of published information is immense, and no researcher can hope to keep up with important developments by merely subscribing to a few journals. Furthermore, the costs of production and distribution have created burdens for publishers and libraries, as well as subscribers. Electronic journals may therefore speed dissemination of ideas and change social practices.

Will the printed journal be replaced by electronic publications? Electronic journals will certainly have advantages: articles can be speedily and cheaply disseminated; subscribers can receive only the articles they want to read, rather than entire bound issues that may contain only one or two papers of interest. If all articles were available electronically, they would constitute a vast, searchable database, the so-called "internet library" or "large-scale digital library." Subscribers could register their research interests and be notified of all new work that might be of interest, receiving only what they have selected. Articles need not be static repositories of outdated thought: as new information becomes available, the original author—or new readers—might add hypertext annotations, perhaps improving a reference section, amending a theory, or adding links to later work.

The technology for a system of electronic journals is largely available, though only about 30 actually exist. The *Directory of Electronic Journals, Newsletters, and Academic Discussion Lists* (Association of Research Libraries, Washington, D.C.) provides bibliographic access to electronic documents and grows larger in each new edition. The impediments may primarily be human: authors publish in traditional journals at least in part for the prestige and recognition that lead to professional advancement. As authors of electronic papers, they might lose professional standing, at least temporarily. Yet electronic journals would provide convenience and thoroughness for scientists and engineers in their research. Measures of prestige will inevitably evolve. We may someday consider how frequently a paper is *accessed* as a criterion for tenure, status, and advancement.

Although imagining a world without printed journals is difficult, computers have facilitated dramatic changes in the way journal articles are written, refereed, produced, and distributed. Writing now routinely includes e-mail contact between collaborators. Citations are probably derived from computer searches of appropriate databases. The text is almost certainly produced—and often submitted—electronically. Referee requests and reports are likely to be sent by e-mail or fax. The major limitation on full electronic submission has been that graphics cannot ordinarily be transmitted with the same verisimilitude as text, but that problem is being solved with special reading devices loaded with the text.

Full electronic journals cost more than conventional printed journals, so the printed journal will probably be around for some time. And if the printed journal does become obsolete, the effect on working scientists may be only minimal, as so much reading is of preprints anyway.

12

Instructions and Procedures

An R&D company has just hired a team to develop manufacturing specifications for a new product. As work begins, the company attorney asks to review the laboratory's safety procedures. To her surprise, the manager discovers that no policies have been drafted. The attorney recommends that work stop. An accident under these conditions would expose the company to serious liability. New employees will sit idle until adequate procedures are written and approved.

Science and engineering present numerous occasions for defining operations—in lengthy documents that exist in their own right and also brief, specialized documents that are parts of longer works. Some will be skimmed but rarely read (consider the methods section in a journal article). Some will be read carefully so that readers can perform the described process. Some may have relatively long lives in service: reference guides serve this function. Or, like installation guides, they may be consulted once. Some, like engineering notes, may never be consulted at all.

Instructions are written so that a reader can accomplish something. Procedures, on the other hand, explain how something has already been accomplished. Thus an instruction might tell readers how to work a system, while a procedure would explain how the system works (or *should* work, in the case of specifications or quality assurance testing handbooks).

Yet despite their marked differences in purpose, length, format, and style, these documents have much in common. They all present challenging writing and reading tasks. You must deliver highly detailed and usually tedious content so that the reader can learn from it. To complicate your task, these documents must often be contractual, with significant legal implications.

Including Liability and Product Warnings

Ideal engineering design would result in products and processes that could be understood and manipulated in accord with common sense. At present, however, most products and processes require instructions to mediate between the technology and the user. The manual (or the video-tape or the online tutorial) facilitates installation, use, maintenance, and repair.

Procedural writing has some force in law. Poorly written procedures can cause problems ranging from frustration and costly delays to injury and death. An injured worker who attempted to follow inaccurate or even ambiguous instructions might be able to collect damages for injury. Several liability cases have affirmed the principle that operator's manuals must enable workers to operate equipment safely.

You must clearly and forcefully warn users of all risks and hazards, both with normal use and with possible misuse of the procedure. Correct verbal content is not enough. Warnings and cautions must be placed well in advance of the point they are needed, and they must *look* different from the rest of the text. Many military specifications (milspecs) contain good models for safety warnings (Figure 12.1). The American National Standards Institute (ANSI) has developed verbal and visual guidelines for warnings, including signs, safety symbols, and accident prevention tags. The ANSI catalog is available on the World Wide Web (http://www.ansi.org/).

Considering Your Audience

Research studies in technical communication, educational psychology, document design, human factors engineering, and information science yield at least one uniform result: readers skim instructions to get *their* work done, not to read instructions. They want to find the information they need, and they want to understand the information they find, while spending as little time as possible reading.

Figure 12.1
These standard warning icons stand out from text and are both dramatic and readable.

When readers are learning to do something new, like use a spreadsheet program, they prefer instructions that give them less to read and more to do. They prefer tutorials that give them a chance to practice and accomplish real tasks. They prefer not to be stuffed with information. Many readers learn better from texts that have less to read and more to look at. If they come to the task with some prior knowledge, they may be able to complete their work without reading the text, learning only from photographs or line drawings.

Audience analysis is always a central task in document planning. In most cases, you discover that you must address multiple audiences with varied reasons for using your document. In writing instructions, you need also to analyze the process itself by sorting it into steps. Such a task analysis—sometimes best accomplished by working the process out in rough flowchart fashion—provides feedback on how potential audiences might behave.

Organizing Your Document

When you have pictured the readers of your document, you are better able to organize information to be most helpful to your audience. Remember that while your problem in writing is to decide how you will place and store information, the reader's problem will be to retrieve what you have put there. In what order should you introduce the steps of a process so that readers can follow and learn? Unfortunately, no single answer will work for all readers. Two strategies, however, will at least bridge the gap between your sense of how someone might best learn and the learner's own needs and goals.

First, we recommend that you make your organization visible and explicit. You know what it is, but your reader does not. Reduce the learning burden by explaining how your document works and how you expect readers to learn from it.

Second, provide an alternative path for users who want to create their own information trails. Rather than punishing readers who do not want to follow you step by step through a process, make it possible for them to learn on their own. Some of your readers will choose linear access to your document, following your instructions as you have anticipated. Others will choose random access. Using the index, they may jump directly to areas of concern, reading only headings, looking only at illustrations (Figure 12.2). A successful document will enable readers to find what they need in the time they are willing to spend. As a writer, you will need to be flexible, creating a document that accommodates more than one style of reading and learning.

If you are selecting an organization based on your analysis of the audience and the constraints presented by the procedure itself, here are some familiar strategies:

- Alphabetical order
- Chronological order
- Cause and effect
- Order of importance
- Spatial order
- Division by task
- Division by component part

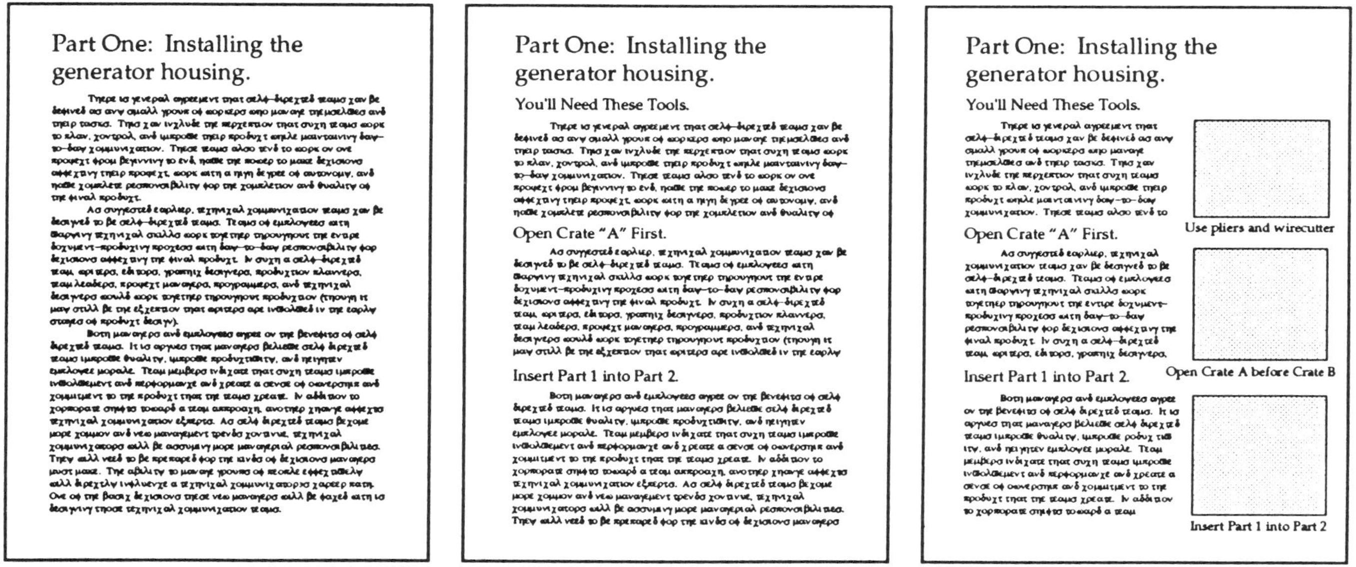

Part One: Installing the
generator housing.

Part One: Installing the
generator housing.

You'll Need These Tools.

Open Crate "A" First.

Insert Part 1 into Part 2

Part One: Installing the
generator housing.

You'll Need These Tools.

Use pliers and wirecutter

Open Crate "A" First.

Insert Part 1 into Part 2

Open Crate A before Crate B

Insert Part 1 into Part 2

Figure 12.2
Of these three pages of procedural instructions, one requires reading solid text. The second adds headings, providing the reader with an alternate path through the document. The third adds illustrations with captions, giving readers the most freedom to learn and accomplish in their own way.

- Database
- Hypertext, multimedia

Alphabetical order is a successful organizational strategy for many documents but a disaster for others. For *learning* a word-processing system, for example, information organized alphabetically would be relatively useless: a novice could not learn from a list of paragraphs about ASCII files, bold type, caps lock key, directories, endnotes, and so forth from A to Z. On the other hand, alphabetical order is extremely useful for reference guides. Users who know a system will prefer speedy alphabetical access to any issue.

Chronological order is a good choice when the steps of a procedure must be followed in order. You might also arrange information by cause and effect or order of importance (simple to complex, increasing to decreasing, most used to least used). Spatial order (left to right, top to bottom) works well when accompanied by illustrations. Division by task and division by component part are patterns that can match function and save readers from skimming an entire document. All of these patterns support a document in which information is stored by the writer in *one way*. With the addition of an index and informational elements like headers and tabs to indicate what material is on a page, a document with information stored in one way can be used in multiple ways.

A database provides massively expanded possibilities for storing and retrieving information. In a database, information is stored in pre-specified fields. Employee files might have one field for names, one for social security numbers, one for addresses, one for years of service, one for salaries, and one for health insurance providers. The information in each category is stored as separate elements and can be retrieved as separate elements: perhaps a list of employees who have chosen health coverage or a list of employees who have been with the company for 10 years or longer. Hypertext and multimedia applications (see chapter 13) provide even greater possibilities for storing and retrieving information.

Achieving Clarity

You can achieve clarity in procedures from a variety of strategies, some verbal, some visual, some organizational:

• Give readers advance information about what they will be reading. Informative overviews and headings have dramatic impact on reading comprehension. Readers learn more when they know what they will be learning. They don't have to spend information-processing time trying to determine the topic.

• Establish a consistent way of naming elements in your procedure and stick to it. Decide whether you will say video display terminal, cathode ray tube, or monitor. Do not vary your choice throughout the document. Consider including a glossary of terms.

• Divide the procedure into modules or segments that allow users to work without turning pages at inconvenient times. Make stopping and restarting easy. If you begin each new module on a new right-facing page or on a new spread of two pages, the physical structure of the document will mirror the modular steps of the procedure.

• Write with verbs that explicitly name the action you want your reader to perform. Figure 12.3 excerpts a very small part of several hundred pages of preferred verbs provided to U.S. defense contractors. This "milspeak" may seem mechanical if you want to be known for a distinctive prose style, but it does enhance clarity.

• Write in the active voice: instead of "the wheel is to be greased," write "grease the wheel."

• Select an appropriate document format. Off-sized pages can be more motivating than the standard $8\frac{1}{2}$- × 11-inch format; spiral bindings permit learners to keep a manual open.

• Illustrate liberally. Remember that many readers learn better from pictures than from text.

• Accommodate random flipping through pages, the search method that most studies show is still many readers' favorite. Headings, highlighting, and illustrations give a user the freedom to search for what's needed in idiosyncratic ways.

• Consider alternatives to conventional linear text. Include, for example, numbered or bulleted lists or message matrixes.

Readability

Technical communicators could use a simple and accurate way to measure the readability of any document. Dozens of readability formulas—some manual, some electronic—have been devised. Most focus on sentence length and complexity of vocabulary as key factors that can be manipulated to improve reading speed and accuracy. The Fog Index (Figure 12.4) is probably the best-known formula.

202 Instructions and Procedures

ML-M-81927(AS)

Preferred Verbs

Verb	Definition	Example
Actuate	To put into mechanical motion or action; to move to action.	Actuate the handpump until pressure gage indicates 50 psi.
Adapt	To make fit a new situation or use, often by modifying.	Use the bushing to adapt the fuse to the projectile.
Add	To put more in.	Add electrolyte to battery.
Advance	To move forward; to move ahead.	Advance the throttle.
Wire	To provide with wire, to use wire on.	Wire the circuit.
Withdraw	To take back, away, or out.	Withdraw the bar magnet from the center of the coil.
Wrap	To wind, coil or twine as to encircle or cover something.	Wrap the wire around the terminal.
Zero	To bring to a desired level or null position	Zero the protractor to the surface

U.S. GOVERNMENT PRINTING OFFICE: 1975-603-131/1427

Figure 12.3
The list of preferred verbs in specifications provided to U.S. defense contractors reduces the creative freedom some writers expect, but it also reduces the reader's learning burden.

COMPUTING THE FOG INDEX

The **Fog Index** uses two factors in measuring readability:

    1.   Average number of words in a sentence (AWS)
    2.   Percentage of words three syllables or longer (%DW)

$0.4 \times (\text{AWS} + \text{\%DW}) = \text{Grade level at which text can be read}$

Figure 12.4
In computing the Fog Index, add the average number of words per sentence to the percentage of long words. Multiply the result by 0.4 for an estimate of the grade level at which the text can be read. (Adapted from Gunning and Kallan 1994. Used with permission.)

Skeptics will point out that readability is an extremely complex issue. We certainly agree: documents acquire readability from a combination of verbal, organizational, and graphic factors, not simply by achieving a numerical score according to a formula. Nor do readability formulas take motivation into account. People will work very hard to interpret extremely difficult prose *if they need to* and have no alternatives.

So even though the results of a readability test are hardly the last word on the clarity of your document, they are still useful. Readability validations are a required stage in complying with many military specifications, and readability software is often supplied with word-processing systems. For example, authors who work on the Athena computer system at MIT can enter the command "style" followed by a filename to receive extensive information about their document, including calculations of readability according to several well-known formulas. Mechanical though such information may be, it does help you revise, particularly in editing excessively long sentences.

Usability

If your document explains how to accomplish something, go through a full rehearsal of the process as you've written it. Give a draft version to prototypical users for walkthrough, testing, and feedback. Such a trial is often called beta testing. There is no better way to gather information about the usefulness of an information product or to find defects when they can be corrected.

Document usability is commonly measured through pre- and post-tests, interviews, observation, questionnaires, and read-aloud protocols in which users read a document aloud and express thoughts about it as they attempt to learn from it. Some usability measures are relatively easy to administer and score; others are both complex to administer and time-consuming to appraise. Though none of these methods has absolute validity, each produces feedback about document function, not just about grammar and style.

Writing Computer Documentation

Most working adults have by now spent numerous hours learning to use computer systems, and many have formed a poor impression of both paper and on-line computer documentation. In the early days, computer product developers stressed feasibility and reliability, not user satisfaction. Products were designed first, and documentation was an add-on. Usability was constructed through the documentation because it was not available in the product. Most users had a hard time learning from that generation of computer manuals. Many found alternative ways to learn what they needed to get their work done.

In recent years, user satisfaction has come to distinguish one computer product from another. Manufacturers now design usability into the product, not just the tutorial. The documentation no longer bears the full burden of making technology available to users: products are less difficult to learn. But computer users now expect *good* documentation, and many technical writing groups have set high standards for manuals and other instructional material.

Documentation Libraries

Most computer documentation requires a library of instructional materials: a first-day tutorial, a task-oriented guide to advanced features, an alphabetically organized reference guide, a template that fits on the keyboard, an on-line reference guide. The first-day tutorial can be book based or online. The library does not have to be large: a laminated, liberally illustrated reference card prepared as an add-on to a printed manual may be more widely used than the document it summarizes.

Task-Oriented Documentation

Effective documentation is task oriented. Its organization mirrors what users are doing. Novice users need a document structure closely allied to work they want to do: headings like "Writing and Editing Your Report" or "Storing and Retrieving Your Document" signal a useful structure. Task analysis is central to preparing helpful information products, and successful documentation is typically the result of extensive task inventories.

Documentation Accuracy

Effective documentation is accurate. It reflects the realities that the user encounters. As the software is modified, the documentation must be modified, and users must receive updates. For some software systems, developers provide documentation on CD-ROM. Information can be conveniently updated and shipped on a new disk, and the entire database can be searched electronically.

Layering

Effective documentation is layered. Users can retrieve beginning levels of information on a subject without also retrieving advanced levels. Layering is an important attribute of both paper and online documentation, though of course much easier to implement online. In some on-line help systems that incorporate expert system features, users identify themselves as they ask the system for information. The system then remembers each request and provides a more comprehensive answer each time the identified person asks for information on that subject.

Accessibility

Effective documentation facilitates reading. The $5\frac{1}{2}$- $\times$ 9-inch printed documentation is physically easier to handle while working at a terminal. Its compressed size automatically cuts the amount of information placed on one page. Computer users learn better from manuals with comprehensive tables of contents, indexes, headings, previews, and summaries. They prefer generous use of white space and markers that make switching attention from page to screen and back possible without losing their place. They do not want to turn pages of manuals to locate crucial illustrations (Figure 12.5).

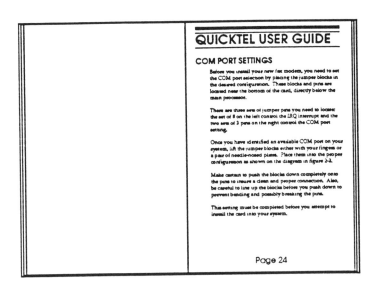

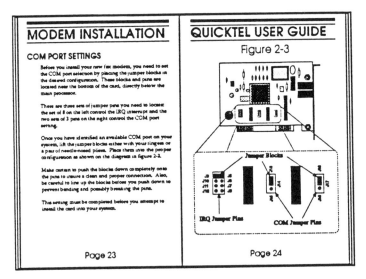

Figure 12.5

These two versions of the Quicktel user guide show the advantage of printing on two sides of each page. In the version at the top, readers must turn pages to find illustrations related to the instructional text. In the second version, readers get verbal and visual information on one spread.

The ISO 9000

The ISO 9000 initiative of the International Organization for Standardization has had a major worldwide impact on procedure writing. To be ISO 9000 certified, a company must document each procedure connected with the production of goods or services. The initial goal of ISO 9000 was to validate consistency and quality so that products could cross borders within the European Community. By 1995, nearly 100 countries, including Japan, had recognized the standards (see http://www.iso.ch).

Though ISO 9000 standards do not specify document formats, many companies (often with the help of consultants) design templates to be used by procedure writers (see chapter 16). Several software packages are available to help multiple authors achieve consistency and clarity. Consistent documentation is increasingly important for global communication and product development.

The Future of Instructions and Procedures

Procedural writing is undergoing dramatic evolution as advances in computer technology make it possible to eliminate the time delays in producing printed materials. For example, since 1985, the U.S. Department of Defense has worked to implement the concept of Computer-Aided Acquisition and Logistics Support (CALS). The CALS initiative provides standards for delivering data in digital form. The result is the "pageless tech manual." Only one copy exists of any document, and users access it electronically. The pageless manual is always accurate, updated as often as necessary, and easy to search.

In moving to electronic text, the time required to create a manual will increase, and so will the initial cost. But the benefits are likely to be substantial as readers receive the information they need, in forms they prefer: viewing a video to see a step being performed, working a simulation to learn an operation, selecting explanatory links to improve understanding.

13

Electronic Text

You've just returned from a long weekend to discover that your company has finally installed the new computer system you proposed almost a year ago. You're delighted, but you also face a challenge. The new system means that most interdepartmental documentation will now be transmitted on-line. Complicated reports and specifications will be stored in the system for easy access, and writers will be presenting information on both the screen and the page. Your task now is to prepare document standards for on-line review. Your writers will need to create printed and on-line help from one source document.

Rather than being bound and fixed on printed pages, electronic "books" are compound documents composed of text, graphics, video images, and audio. These documents can be "intelligent" and "interactive." Their sequence and even their style of presentation can be selected by the reader. An electronic document can build itself by extracting information from a database and send itself to designated recipients. In settings where security is an issue, electronic documents can conceal contents for users with limited clearance.

On-line communication solves a variety of problems associated with paper. Electronic documents can be customized: personal training manuals can be created for each learner and based on the trainee's performance on tests of skill. These documents are easy to update, so they can always be accurate. They are easier to search than books, providing improved access to topics and cross-references. They can be remarkably compact: a laptop with CD-ROM drive can deliver a 10,000-page documentation set that would have occupied a dozen three-ring binders.

Writing an Electronic Document

Electronic documents can supplement paper manuals or replace them, but using a video display terminal to provide information does not guarantee usable documentation. On-line information needs to be structured for the screen. Displayed pages from printed books will rarely yield effective on-line material.

Writers of electronic documents need to create clear specifications for visuals and typography, just as they do for hard copy. They must provide for the unique ways users interact with on-line material, facilitating multiple types and levels of searches. They must structure material to minimize a user's disorientation, providing ways for readers to tell where they are at all times.

Designing for the Computer Screen

Writing electronic text presents real challenges to information designers. People are used to the physical features of books. A certain heft suggests the time it will take to read the text. Page numbers are visible signals of progress. Pages can be marked and dog-eared. Bookmarks can be placed and replaced. Pages are present even when they are not being read.

With on-line text, users have different cognitive challenges. A screen has less space than a standard page, and displayed text is almost always less legible than it is when printed. For most people, moving through several computer screens is not as easy as looking back and forth between pages of a book. On-line information can be confusing: users do not know where they are and where they are going unless orienting markers—numbers and titles for each screen, system maps, and exit buttons—are built in.

Conventions for writing electronic text are evolving. Opinions conflict about which fonts are most legible on screen, which graphic-highlighting devices attract a reader's attention, and what effects color, blinking, sound, and animation will have on reading comprehension. Some relatively uncontroversial techniques for improving on-line text include matters of language, screen design, and user psychology:

• Write concisely, with only small chunks of text to read on each screen. Conventional wisdom holds that readers can deal with seven (plus or minus two) pieces of information. On computer screen, it appears that a standard of five (plus or minus one) works better to minimize confusion.

• Use simple language so that reader get it right the first time. Most users of on-line documents do now want to relocate and reread anything!

• Design all screens in the same format and typographical style. Use a limited number of fonts, styles, and colors. Select fonts that are particularly legible on-screen. Except for headings, use upper- and lowercase letters.

• Allow users to choose sequences in which they will read the text. Build in methods for users to sign off and later return to the exact spot they left. Provide navigation and escape information on every screen.

Hypertext and Multimedia

Traditional communication is linear. Information is laid out in a single path. Hypertext, in contrast, is composed of individual chunks of content and computer-supported links among these chunks. Readers follow topics in any order they choose, sometimes guided by a map of the network, sometimes creating their own paths. The chunks of content can include sentences, paragraphs, definitions, annotations, and drawings. In multimedia systems, the chunks can include sound and video.

Multimedia documents present new possibilities for information storage and retrieval: not just linear to random access but linear to multidimensional. Each unit can be electronically linked to any other unit, and the user can choose which moves to make. At every step, the user of a multimedia system can see an example or a simulation, look up a definition, listen to sounds, return to a previous link, or create links. Rather than a fixed order of presentation, users have rich opportunities to travel through what has been called a *docuverse*, a universe of documents. Depending on needs and interests, each user can take a distinctive route through complex material.

A well-designed multimedia system organizes data in a complex, non-linear way and facilitates exploration of large bodies of knowledge.

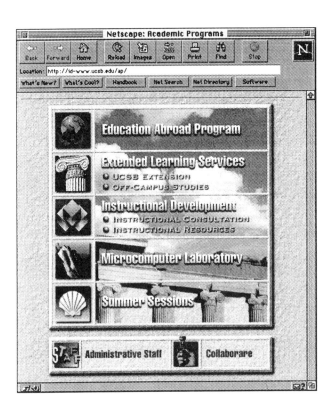

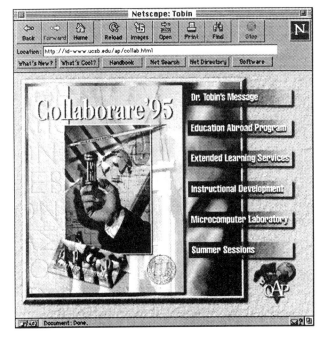

Such a system works when many users need varied levels of access to large amounts of loosely structured information. These applications are, obviously, suited only to environments in which computers are always available.

You can produce two kinds of hyperdocuments: one kind is converted from previously written work to hypertext format; the other is explicitly created for hypertext. Both are challenging to build and not always easy to use. The usefulness of hypertext systems depends on how well the information has been chunked and linked by writers and how flexibly it can be accessed by users. Designers must know which chunks make sense, which links are required.

Writing for the World Wide Web

The World Wide Web provides a platform-independent way for text, images, sounds, animation, and video to be "read" by numerous readers at remote sites. Addresses on the Web are called URLs (Uniform Resource Locators). The language of the Web is HTML, and several software packages convert word-processed documents to HTML format. Existing graphics can be converted to standard Web graphic formats (GIF files and JPEG files), and multilayered, multimedia Web documents can be created almost overnight.

Many of the principles that guide the production of printed documents apply to the development of Web pages. But Web documents are layers deep—not pages long—and planning a Web site calls for an "information architecture" mindset. In a good Web document, readers can move efficiently from one topic to another (Figure 13.1). Many Web sites are interactive: visitors can communicate with Webmasters and other site visitors by way of preformed e-mail templates.

Figure 13.1
The home page of the University of California, Santa Barbara, Office of Academic Programs (top) provides links to a variety of topics. Selection of the topic *Collabore* takes the Netscape voyager to a set of six options for further investigation (bottom). (Courtesy of Steven Brown, Principal Artist, University of California, Santa Barbara)

A Web designer must create linked connections as well as chunks of content. But self-sufficient chunks or modules of information are not enough on the Web. Granularity—the relationship between content chunks and physical breaks on the screen—is also a crucial consideration. Audience analysis is a tricky task in this medium: what do you want visitors to learn from your site? what links will they need? will graphics, sound, or animation improve or impede learning?

One of the most efficient ways to derive principles of effective Web page design is to explore existing sites with a critical eye to what works for you and what doesn't, with the goal of preparing a style guide for your writing and design. Three style guide sites are good models:

• Yale   University   (http://info.med.yale.edu/caim/StyleManual_Top. HTML)
• World Wide Web Consortium (http://www.w3.org/pub/WWW/)
• Sun   Microsystems   (http://www.sun.com/styleguide/tables/Welcome. html)

Past and Future Applications

Multimedia has been with us at least since 1978, when the Architecture Machine Group at MIT developed the Aspen Movie Map. This was a "surrogate travel" application that allowed the user to take a simulated drive through the city of Aspen. A set of videodisks contained photographs of all the streets in the city and some of the buildings. Users could stop in front of many buildings and go inside! The Aspen Movie Map even had a time-of-year knob, giving the user a choice of the autumn or the winter version.

The Architecture Machine Group also created a prototype Movie Manual, suitable for both novice and expert auto mechanics. Among the features of the Movie Manual was this one: each time a tool was mentioned, the mechanic could link to a picture of the tool and a narrative about how it is used *or* to a video of an experienced mechanic using that tool.

A newer application is RED, now in tests by the United States Army. RED is a portable computer system that delivers spoken information and screen images to technicians in the field. The user wears a belt, earphones, and a small screen attached to headgear. The belt contains a computer that plays CDs capable of holding the equivalent of 550,000 pages of text. The earphones deliver spoken information about performing field tests; the screen produces images, allowing the user to do the maintenance tasks without turning pages of a manual. The RED configuration can be personalized for each user, with a "smart card" program that provides a checklist of tasks that the worker is to perform on a particular day. Multimedia shows promise for a great many similar applications.

A Paperless Future?

Enthusiasts are convinced that hypertext and hypermedia applications will be the basis of a new literacy. Software will diagnose the user's abilities and learning needs, and the multimedia "book" will reconfigure to best suit each "reader." Learning will be effective and powerful because, in this view, hypertext systems model the associative style of human idea processing. Information will always be timely, because electronic updates are cheap and convenient.

Skeptics wonder about a future in which all texts are unstable and can be read in any order, perhaps revised by many readers. Which, if any, versions of a document will be authoritative? What factors of electronic text will substitute for the social signals that distinguish a high-quality printed book from a carelessly prepared handout? What do people need to learn so that they can browse profitably in immense multimedia databases? What will be the long-term effects of non-linear, multimedia "reading"? What about the issue of intellectual property in easily reproduced electronic documents? Can electronic documents be censored? What is the longevity of digital information? Will electronic documents become obsolete when hardware and software change?

We are in a dramatic transition from paper to on-line documents. Why provide each of 600 employees with a 500-page manual that needs updates at least twice a year? Accurate and updated information can be delivered on replaceable CD-ROMs, on the Internet, or on a proprietary Intranet.

Readers are more familiar with paper-based than on-line formats, and they do not always know how to learn from electronic documents. Electronic text is less legible, and slower and more tiring to read. Multimedia documents are more time consuming and expensive to produce. But the potential advantages of electronic documents include vast storage capacity, easy search and retrieval, accuracy, and economy.

14

Oral Presentations

Your project is nearly finished, and you're pleased with the design and possible applications your group has considered. Now, as you enter the development phase, you will be assembling a new team. Your next move seems obvious: to present aspects of your work at a professional meeting. The exposure for your research will help attract colleagues who might become part of the project. A formal presentation should also enhance your stature as a research scientist. But first, you need to make your presentation clear and engaging.

The life of a project presents numerous occasions to talk: in relatively informal group problem-solving sessions, in briefings or question-and-answer sessions with clients and sponsors, in formal presentations at professional meetings. On these occasions, you are most likely to be in the same room with your listeners, although technology may eventually make physical distance between speaker and audience irrelevant.

Engineers and scientists often approach oral presentations with anxiety. Perhaps they should. Talks can be hard, even impossible to follow. Slides or overhead transparencies can be too complex to read in the few minutes they are displayed. Talks scheduled for 10-minute time slots sometimes go on for 15 or 20 minutes, in blatant disregard of the other scheduled speakers and of the audience. Surely no presenter sets out to be boring, obscure, and insensitive, but unfortunately, many are.

No one method guarantees transfer of knowledge directly from your head to the heads of your listeners. You may want to mimic the medieval German woodcut, in which a schoolmaster uses the legendary Funnel of Nürnberg to pour learning directly from a bottle into his student's head.

Unfortunately, for technical communicators, there is no Nürnberg Funnel and not much possibility of finding one. Preparing and presenting technical information that reaches listeners means considering factors that are not always easy to assess. You do know, from your own experience as an ear- and eyewitness to unsuccessful presentations, that getting all the words and numbers in is not enough.

Nor is it enough (or even necessary) to have the slick delivery style of a network news announcer or to have professionally prepared multimedia props. Well-constructed content is the crucial factor in presentation success: you have a strong and interesting idea, and you make the presentation fit the listening and learning styles of your audience.

Most of us can remember times when we could not learn from talks because the speaker was incomprehensible. We can probably also remember speakers—often professors—who violated our every expectation about successful communication and yet were comprehensible and even inspiring. Listeners are interested in ideas and techniques that they can take from the talk and apply to their own work. When they attend a talk at which valuable ideas are put forward, they are remarkably forgiving of less-than-ideal delivery style.

Listening versus Reading

As difficult as many written documents may be, they are still potentially easier to learn from than oral presentations. Readers can choose their pace and return to sections that demand further study. Listeners cannot go back over what they have just heard, and they cannot ask the speaker to stop so that they can follow a thought of their own. In preparing oral presentations, you need approaches different from those for documents —and the differences apply to the visuals as well as to the text.

A talk should not be a "speechified" report or journal article. Your presentation must conform to a different model of information transfer. Accommodate your subject to the way people learn from listening: stick to a few main points rather than to every line of thought connected with your subject. Remember that the audience for an oral presentation is often less specialized than the audience for a journal article. If the overhead transparency with the outline of your talk reproduces the table of contents of your paper, you may have neglected to prune and shape your topic for the benefit of listeners. In Figure 14.1, the generic table of contents at the top is suitable for a written report but not for an oral presentation. Displayed underneath is a suitable table of contents for the first overhead in the oral presentation.

Each January, Professor Patrick Winston, one of the founders of the Artificial Intelligence (AI) Laboratory at MIT, gives a presentation called "Some Lecturing Heuristics" to overflow crowds of students and faculty. Winston applies the AI concept of frames to the construction of talks from which listeners can learn. A frame is a structure for containing data, a blank to be filled in, like a name and address box on an application form. People are likely to come to your talk with some frames ready to be filled about your content and focus. They will then base their expectations on your title and abstract and perhaps your previous work.

These generic frames will be modified by a specific lecture. You can facilitate learning by explicitly providing listeners with frames: spaces where concepts can be entered. Start with the general picture, Winston advises. Put an outline on the board; tell the audience what frames you will be filling in, so they will not have to figure out what you are talking about and how your talk is organized. Be explicit: say things like "By the time we're finished here today, I want to convince you. . . ." Your talk can then recall frames, fill them in, link them together, and in some cases, change them, or even create new ones.

Your Audience and Environment

To develop an effective presentation, you need to combine what you know about how people learn from listening with what you know about your particular audience. Who are they? How many of them will you face? Why have they come to hear your talk? What do they already know about your subject?

You also need to consider the physical and social environment in which you will speak. What kind of room? What time of day? What has the audience been doing before your talk? What will participants do afterward? Are you the only speaker, or are you competing for time and attention as member of a panel?

Desalination of Salt Water
Using Wind Energy

Table of Contents

iii

Desalination of Salt Water
Using Wind Energy

☑ Fresh drinking water is an endangered resource.

- •U.S. consumption 360 billion gallons/day
- •Severe depletion of aquifers

☑ Wind energy can be used to desalinate salt water.

- •Important sources of salt water
- •Desalination can utilize off-peak electricity

☑ Technology can be applied at numerous U.S. and international sites.

The answers to these questions will not always be what you might wish, but you will not be able to keep secrets from your audience. The room may be dark, airless, and poorly soundproofed. The time for your talk may be Sunday morning at 8:30 A.M., on the last day of a 4-day meeting, or it may be the last time slot before lunch break or the first time slot after lunch! Audiences respond positively to speakers who acknowledge mutual concerns and often with hostility to speakers who do not. If you carry on as though nothing is wrong while your talk is nearly drowned out by the noise from rooms on either side, audience members may well wonder if you care about what is happening to them. In a case like this, you might indicate that you, like them, find this setting unacceptable and that you hope to have informal conversations on your subject with members of the audience at another time, in a more suitable place.

Structuring Your Talk

In presentations, your aim should be to *uncover* key points rather than to *cover* every detail. Your talk is probably not the first word on any subject, and it does not need to be the last. A written version may be available to your audience in the conference proceedings or as a handout. Organize your talk for listeners, not readers. Ask yourself, Given what I know about my audience, what is the clearest and most convincing sequence in which to order the information in my talk?

When you have developed a strong organization, be sure to make it obvious and explicit. The well-known preacher's wisdom is good advice: "Tell 'em what you're going to say. Say it. Tell 'em what you said." In successful presentations, you need to develop the technical content of your talk, but you need also to develop the *meta*-content, the verbal and visual structures that allow listeners to learn from your talk.

Figure 14.1
The table of contents at the top is perfectly adequate for a written report. Yet when the author prepared a table of contents for his oral presentation, he created a new set of descriptors for the overhead transparency. These are displayed below. Note that the language for the overhead is informative, not generic. Three key points are covered, and the overhead creates frames for each point.

Technology,
Networks,
and

**THE LIBRARY
OF THE FUTURE**

Jerome H. Saltzer
Prof. of Computer Science
M.I.T.
Saltzer@MIT.EDU

Overview

1. The vision
2. The driving technology
3. The system engineering challenges

The Enabling Technologies

- High-res desktop display

- MB/Sec data communications

- VERY big disks

- Client/server architecture

Space Needed
(300 pixels/inch, compressed)

1 400-page book	30 Megabytes
2-M books (M.I.T.)	60 Terabytes
60-M books (L/C)	1.8 Petabytes

$$\text{Tera} = 10^{12}$$
$$\text{Peta} = 10^{15}$$
$$\text{Exa} = 10^{18}$$

Challenges

- Applying Client/Server Design

- Links

- Persistence

Conclusions

1. A revolution is available for the asking.

2. The engineering is a fascinating challenge.

Begin technical presentations with explicit discussion of the way your talk is organized. Preview main points. Define specialized terms or key phrases. Tell listeners when you are finishing one section of your talk and starting another ("I've talked about cost factors, and now I want to turn my attention to environmental concerns"). End technical presentations with a review of main points, saving your best formulations for last so that listeners will not be irritated at a mechanical reiteration of what they have just heard.

Integrating Visuals

Because listeners learn and follow better when they have something to look at in addition to something to listen to, visual props are a standard feature of technical talks. Even if you think that visuals are probably not important for understanding your subject, everyone else will probably bring laptop computers, trays of slides, or packets of overhead transparencies. In contrast, your talk may seem underprepared. The overheads reproduced in Figure 14.2 are skillfully designed to help listeners follow an argument about the library of the future, a library in which 2 million books can be stored digitally in 300 square feet.

Plan the content and progression of visuals *as you plan the organization of your talk*—not as a separate process. Many technical professionals use the moviemaker's technique of storyboarding to plan and prepare presentations. Text and visuals are then usually successfully integrated. Software is now available to support the storyboarding process, but a do-it-yourself version works well. For a 15-minute talk, prepare an outline and then work with at least 12 pieces of $8\frac{1}{2}$- × 11-inch paper in horizontal orientation. Each sheet represents both the text and the visuals for each minute or so of the talk. Plan the first two sheets as word

Figure 14.2
These overheads were prepared by Professor Jerry Saltzer of MIT. The first overhead named the subject, and the next, called "Overview," previewed the order in which the subject would be treated. Additional overheads, combining text and drawing, provided various kinds of support for listeners, including repetition of keywords and verbal and visual definitions. Note that the last overhead is explicitly labeled "Conclusions."

charts: the first should display the title of the talk and your name; the second should display the overall outline of your presentation. For the remainder, divide each page in two. Use the left side to jot down rough notes for the narration and the right side to sketch the content of the accompanying visual. Tack or tape the completed storyboards to a wall so that you can preview the entire presentation as a unit. With this global preview, you can search for duplications and omissions. You can insert word charts to mark each transition in your talk. You can create a memorable concluding visual. This technique provides better feedback than separate assessment of text and visuals. It more nearly models the two streams of information that your audience will be receiving.

Selecting a Visual Medium

In deciding which visual medium is most appropriate, you will need to apply what you know about the audience and environment for your presentation. What visuals do these listeners expect? What visuals will other presenters be using? What visuals will work well in the room? Answer these questions before you determine what is technically possible. A multiprojector, multimedia presentation is not necessarily well-suited to all subjects, audiences, or settings.

Chalkboards Despite its limitations, the chalkboard or markerboard remains a favorite medium, particularly for in-house or academic presenters. It is widely available, and it does not require training or electricity. Because copy is not fixed, the speaker can operate on it: adding, deleting, circling, underlining, color highlighting. Though writing on a board slows a presentation, the delay can also give listeners time to absorb what you are saying. And you can, in some cases, write or draw on the board before your presentation.

On the downside, board space is limited. Rearranging items is time-consuming when they must be erased and redrawn, and items disappear when the space is needed for another topic. Even chalkboards are evolving these days toward electronic forms that combine the features of the traditional chalkboard with the storage and communication links of workstations. The key to working in these new media is still the notes that oral presenters have prepared for structuring their talks.

Overhead Transparencies Overhead transparencies are, for good reason, very widely used to support technical presentations. Unlike slides, which require a darkened room, overheads can be viewed in ordinary room light. They are inexpensive and easy to prepare. They are easy to carry and store, and they can be retrieved during a question-and-answer session.

Overheads are a dynamic medium: speakers can write on them to emphasize points, cover part of the transparency to progressively disclose the image, and prepare overlay transparencies to be placed over the first. At some presentations, attendees are given a photocopied set of overhead transparencies. When overheads or slides are prepared with presentation software packages, you can present the audience with miniaturized copies of each visual.

Slides If you are speaking in a large room, consider slides. Compared with overheads, slides offer more interesting design possibilities and have a higher visual impact, with outstanding color contrast and accuracy. Many presenters now use standard office software programs like Microsoft's Powerpoint to create computer-based slide shows. Presenters can easily incorporate data from existing files—including spreadsheets, tables, and figures—for a new presentation.

Electronic Presentations Electronic presentations, in which computer-generated images are displayed directly on one or more monitor screens or projected by way of a light valve onto one large screen, provide opportunities for dynamic and interactive meetings. Images are often not only displayed but also manipulated, with input from audience as well as presenter.

Producing Effective Visuals Whatever visual medium you choose to support your presentation, assume that you will need to create new copy. Illustrations transferred directly from a technical report to a presentation slide or overhead transparency are nearly always dense and hard to follow.

Several graphics software packages give users the tools to create images that can be saved as transparent sheets for overhead projectors,

35-mm slides, paper handouts, or a slide show for display on the computer screen. With a film recorder that accepts graphics input directly from a computer, any image generated on the computer screen can be automatically processed and mounted as a slide, eliminating the traditional time lag in producing slides for presentations. Some speakers create presentations in Hypertext Markup Language (HTML). They store the HTML presentations on their own files and access them through the Internet for presentation elsewhere.

Keep in mind these guidelines:

• *Keep visual aids simple and legible.*   Develop one idea per visual—a single point, relationship, or conclusion—with plenty of blank space. If you need to show details, prepare separate visuals on the same subject, progressively disclosing complexity.

• *Design the right number of visuals.*   A workable formula is that the number of overheads or slides should equal two-thirds the number of planned presentation minutes. Of course, the optimal number of visuals depends on subject and audience, but if you have fewer than one-third the number of visuals to the number of minutes, you are probably trying to put too much information on some of your visuals. Conversely, if you move in the other direction, nearer one visual per minute, some of the visuals are probably not staying on the screen long enough—an effect that can be cumulatively irritating to your audience.

Rehearsing Your Talk

When you practice your talk, try putting yourself in the audience's place. Consider how the audience can learn from what you are saying and what you are showing. Consider how much time audience members will need to read and learn from each visual. Don't remove a transparency so quickly that it can't be read. Don't block your own visuals as you advance them. Use a pointer tool, which is less distracting than the shadow of a presenter's finger. Remove visuals you are no longer talking about, and turn off equipment you are no longer using.

Prepare and practice every element of your presentation. The standard instructional tool in workshops to improve presentation style is videotape, enabling you to see and hear yourself as others do. But you can learn quite a lot without a video preview. Time yourself: if the talk is too long, cut before you present it, not as you are giving it.

Rehearse with visuals: practice board work, manipulating transparency overlays, and retrieving slides. Look critically at your overheads or slides from a position in the back of the room. No audience has ever been glad to hear a speaker say, "I know you can't see these slides, but what they show is...." If you feel you have no option but to display a visual that can't be seen, provide the audience with a photocopy to accompany your discussion.

Should you ever read your presentation? Practice will vary, and at academic conferences, some speakers read their papers. The preferred presentation style, however, is well-prepared but conversational, prompted *and* accompanied by overheads or slides, perhaps cued by note cards, not $8\frac{1}{2}$- × 11-inch manuscript sheets. If you do need to read your paper, start out by *not* reading: talk directly to the audience for the first few minutes, setting a context for your paper, adjusting your opening remarks to fit the situation.

Prepare to handle feedback. In question-and-answer sessions, repeat the question so that everyone can hear it. Audiences are frustrated when listening to answers without having heard the question. Do not over-respond; you can offer to continue specialized conversations at another time.

Conference Presentations

Three specialized forms of technical communication are associated with talks given at professional meetings: presentation abstracts, poster sessions, and papers written for conference proceedings.

Presentation Abstracts

Conference organizers frequently ask potential participants to submit a presentation abstract, essentially a proposal to give an oral presentation. Presentation abstracts are usually written many months before the meeting, and they often describe work you have not completed. Still, they need to be informative, detailed, and as complete as you can make them. Some professional societies provide preprinted forms on which to submit presentation abstracts. Abstracts of accepted papers are frequently published and distributed to all meeting registrants. The audience for your abstract may be large, including both those who do not attend the talk and those who do.

Poster Sessions

In poster sessions, speakers are given a bulletin board on which to display graphics and text for a specified period, perhaps 2 hours (Figure 14.3). Poster sessions are often lively and productive discussions between the presenter and a small, interested audience.

Poster session abstracts have their own conventions. The American Association for the Advancement of Science, for example, requires that a poster session abstract fit within a 5-inch square. Above the square, the name of the broad discipline that encompasses the subject matter and three index words are to be provided. If the abstract is accepted, the presenter has 90 minutes to display the poster.

Proceedings Papers

Some conference organizers publish proceedings containing copies of papers given at the meeting. The proceedings may be distributed at the meeting, with text of papers that have not yet been delivered, or they may be published at a later time, with the text presumably revised in light of feedback and discussion. Some proceedings are produced and distributed by the publishing industry, others by conference organizers. Except when they are published after the meeting by an established technical publisher, papers in proceedings are not usually refereed or edited.

Papers in proceedings tend to be shorter than papers in journal articles, not fully developed, and sketchily documented. Journal articles, subjected to rigorous peer review, including textual editing, are more polished. Many speakers who have written papers for conference proceedings later rewrite their findings for submission to a refereed journal.

Science and engineering librarians call conference literature gray literature (*graue Literatur*), because it is often distributed in an unconventional way and therefore difficult to locate. This is not a denigration of the form: in some new fields, much of what is known is available only in conference literature. Conference proceedings are covered in *Engineering Index*, *Science Citation Index*, *Biological Abstracts*, *Chemical Abstracts*, as well as in specialized publications like *Directory of Published Pro-*

Tidal Mini-Hydro Power Plant

Brian Kato

Abstract

Tidal Hydroelectric Power Plants utilize the cycling of tides to power a turbine. These low-head turbines convert the mechanical energy of the water pushing against the turbine blades into electrical energy in a generator. Tidal power generators have many advantages. Tidal power's main advantage is its low cost, due primarily to low maintenance requirements and an absence of fuel costs.

Intro

Tidal power uses minimal land because the majority of the land required is already subject to tidal flood. Tidal power is a renewable resource that produces energy with low environmental impact. Our limited fossil fuel resources are becoming increasingly scarce, and we must search for new and better sources of energy if we are to maintain or improve our current standard of living in this country.

Methods

A reduced-size, fully operational test model of a Tidal Hydro-electric Power Plant was constructed by our design and engineering team to evaluate the structural, cost, and design efficiencies of this system if it were to be implemented on a full-scale basis. Computer-generated modeling, along with a scale replica and numerous sub-assemblies, was used to calculate the actual data reported here. This system was compared with various other alternative energy sources, including wind energy, photovoltaic cells, ocean thermal-energy conversion, and concentrated solar applications. Oil- and natural gas-fired power plants were used as a benchmark for comparison on an economic efficiency level. Consideration must be given to the current economic climate and the rise in cost of these fossil resources, improving the overall cost structure of Tidal Hydro-electric generation systems.

Results

Our model generated $2.56 * 10^{12}$ watts/meter^2 over a period of 9 months after it was put into continuous operation. While we were unable to attain the economic efficiencies of scale that would result from the construction of a full-size system, our costs were surprisingly minimal. The system produced a large percentage of its power when the moon was full, because of tidal cycles. These factors should be considered when implementing the system on a large-scale basis.

Fig. 1

Table 1

Date	Action
January	Design model
March	Construction
May	System testing
June	Data collection
August	On-line model
September	Results

Table 2

Variation	Results
Hydroelectric	123
Solar Cell	456
Nuclear	789
Wind Energy	012
Biomass	345
Passive Solar	678

Conclusions

When compared with various other alternative energy sources, including wind energy, photovoltaic cells, ocean thermal-energy conversion, biomass systems, and concentrated solar applications, as well as oil- and natural gas-fired power plants, tidal-driven hydroelectric generation systems prove to be highly efficient, as well as economically viable. Very low maintenance costs combined with almost no environmental impact means that these systems could be fully implemented by the middle of this decade. In the immediate future, tidal hydroelectric generation can reduce public and private sector reliance on main-grid power by up to 20%, reaching an attainable maximum of 65% by 1998.

Figure 14.3

This poster is based on a physically enlarged, condensed text of the paper as it might be published in a journal. It includes abbreviated versions of standard parts of the written report.

ceedings (InterDok), *Index of Conference Proceedings* (The British Library), and *Index to Scientific and Technical Proceedings* (Institute for Scientific Information).

Telepresence for Meetings of the Future

Improvements in communication technology will make it increasingly feasible to have real-time meetings without the simultaneous presence of participants. In these multimedia settings, new kinds of computer-mediated interactions will be possible, with participants sharing an audio and a visual space. The technical infrastructure of the new conference room will support different models of information transfer—and perhaps better learning. Audiences may be offered less passive roles, and new technologies will change the way people share their ideas.

15

Job Search Documents

Late on a Friday, looking forward to the weekend, you check your e-mail and find an alarming message. You discover that your plans will have to change. The new department manager wants to interview all mid-level staff, and she wants an up-to-date curriculum vitae (CV) by Monday morning. Unfortunately for you, your CV is 7 years out of date. You aren't afraid of losing your job, but to be ready for Monday, you will need not only to revise your copy, listing what you've done for the past 7 years, but also to change your page design. You've discovered what scientists and engineers know and often ignore: that an accurate and attractive CV is a crucial document for professional advancement.

The Formal Professional Biography

As an ongoing task, you should write and update a biography that presents personal data in either a curriculum vitae or a resume. Though the terms are sometimes used interchangeably, a CV is a record of academic, professional, and sometimes personal achievements, while a resume also includes an employment objective.

Except when you are actively engaged in job searching, a CV will be far more useful than a resume. Your CV contains the kind of information that conference chairs want as they introduce you to an audience and that funding agencies want when you apply for support. Your employer also may want to see your CV as you undergo a personnel review.

A good CV or resume relates your strengths and achievements to your purpose—professional review or job search. Your document needs to be drafted in the clearest language, with the most attractive and functional design. Many personnel officers claim that resumes have less than 1 minute to make the right impression. Deliberate and consistent use of page design elements like white space, bullets, italics, and bold type can assure that your strengths are apparent even to readers who do not spend much time studying your resume.

Build your CV or resume of component parts—modules that can be formed and re-formed to map your strengths and achievements. Within each module, present the most recent information first. Which elements should you use? In what order? Let your own achievements be your guide. For some occasions, you may want to emphasize details of academic training. In other cases, you may want to emphasize experience, giving that module a more prominent position on the resume. Match the concerns of your audience and the purpose of the document.

Some resume elements are conventional, but only the first of the following is absolutely required:

- Name, address, telephone number, e-mail address
- Objective
- Educational history
- Employment history
- Special skills and training
- Honors and awards
- Memberships
- Publications
- Conference presentations
- Personal background

Ask several colleagues to read your draft CV or resume, looking for obscure presentation of information. You know what the U of O is, while your reader may wonder if it is the state university of Ohio, Oklahoma, or Oregon. Combine miscellaneous jobs into one category, and don't inadvertently emphasize the insignificant.

How long should a CV or resume be? Many organizations are strongly committed to the one-page resume, even for senior professionals who could fill pages with their achievements. For others, a multiple-page document is acceptable. Find out what you can about what is expected, but assume that you have limited space to work with. A CV or resume should be as short as possible, and you will want to get the maximum payoff from every line.

Keep your CV current. It should reflect the changes in your life, and you should have a procedure for making changes, for dropping some data and adding others. Because a CV should be brief, you will need to weed out older achievements as you add more recent ones. As you move forward, detailing more recent job descriptions, you will take fewer steps backward, inevitably dropping one or more older pieces of your life to make room for the new (Figure 15.1).

Computer files are particularly good ways to keep your biography current. You can have a file of your most recent CV and a record of each version you have prepared. You can also keep an ongoing file of material to incorporate into the next version you prepare, together with names and addresses of actual and potential references.

Application Letters

A letter offering yourself for employment should be brief but detailed and informative. Except in unusual circumstances, it should be written *to* someone. If you want the job, do your best to find the name of an appropriate recipient. The letter of application in Figure 15.2 indicates that the writer has acquainted himself with the company's research agenda. He elaborates on the enclosed resume, providing details that will interest the potential employer. The letter is friendly without exaggeration or insincerity.

Writing letters like this is hard work. It is certainly easier to write a brief note asking "To Whom It May Concern" to study your enclosed resume and decide which of your achicvements would match the company's needs. But the burden of connecting what you have to offer with what the employer requires is on you, the writer. You want something from your reader (a job!), and you should make it easy to see what distinguishes you from other applicants.

The envelope in which you enclose your letter and resume can be a significant part of the application. Because the envelope is postmarked and date-stamped, it frequently remains a part of your application package. You can make the package more attractive by using an envelope

Michael L. Dickens
233 Green Street
Cambridge, MA 02139

Objective   A summer job in computer science or electrical enginering with the potential of full-time employment after graduation.

Languages   Computer: BASIC, Pascal, C, LISP/Scheme, 80286, 68020 Assembly
Foreign: French, German

Education   Massachusetts Institute of Technology   Cambridge, MA

Experience,   MIT ELECTRICAL ENGINEERING DEPARTMENT   Cambridge, MA
Fall   Worked as a teaching assistant for Acoustics under Prof. A. Bose, member of the
Semester,   Electrical Engineering Department. This position entailed regularly scheduled tutoring
1989   sessions and grading of students' work.

Summer,   OAK RIDGE NATIONAL LABORATORIES   Oak Ridge, TN
1989   Employed as a junior summer student in the Engineering, Mathematics, and Physics
Division under Dr. C. Weisbin.

Spring   MIT ELECTRICAL ENGINEERING DEPARTMENT   Cambridge, MA
Semester,   Worked as a lab assistant to students taking course 6.003 Signals and Systems,
1989   instructed by Prof. W. Siebert. Required a knowledge of the HP Bobcat computer,
LISP, and a general understanding of signals and systems.

Fall, 1988   MIT ELECTRICAL ENGINEERING DEPARTMENT   Cambridge, MA
Semester   Worked as a teaching assistant for Acoustics under Professor A. Bose, member of the
Electrical Engineering Department. This position entailed regularly scheduled tutoring
sessions and grading of students' work.

Summer,   OAK RIDGE NATIONAL LABORATORIES   Oak Ridge, TN
1988   Employed as a junior summer student in the Engineering, Mathematics, and Physics
Division under Dr. C. Glover. Contracted by the Air Force to develop a procedure that
graphically represented the breakdown of a program for execution on a parallel
machine. This project prduced a paper that appeared in the Proceedings of the Fourth
Conference on Hypercubes, Concurrent Computers, and Applications, March 1989.

Summer,   Employed as a sophomore summer student in the Engineering, Mathematics, and
1987   Physics Division under Dr. Glover. Took many serial-linked programs and developed
one large parallel program for the VERAC Corporation, an Air Force contractor.

Interests   Letter in lightweight varsity crew. Sports include biking and swimming. Play bass and
piano.

(a)

Figure 15.1
In 1988, Michael Dickens prepared a resume in hopes of getting a summer job. When he revised the document early in 1991, he produced a CV, this time accounting for significant academic and professional achievements. Note that in

Michael L. Dickens
Mailing Address
223 Green Street
Cambridge, MA 02139
(617) 547-0965
E-Mail Address
MLK@ATHENA.MIT.EDU

Education   Massachusetts Institute of Technology   Cambridge, MA
B.S. in Electrical Engineering, February 1991
Courses include Technical Communication, Acoustics, Musical Acoustics, Psychoacoustics,
Speech Lab, Strobe Lab, and Robot Manipulation

Computer Languages: 68020 and 80286 Assembly, PostScript, C, LISP, Pascal, Fortran
Operating Systems: DOX, UNIX, VAX VMS, Macintosh

Experience   **MIT Undergraduate Research Opportunity Program**   Cambridge, MA
Summer 1990: Produced a system for modelling of acoustic tubes with varying cross-sectional
area. Compared modern theories of sound propagation, lossy transmissions, and conically and
exponentially varying structures. The system was programmed in C on a Sun workstation,
involves numerous structures, and includes a routine for complex bessel computation.

MIT Department of Electrical Engineering   Cambridge, MA
Fall 1988 to Spring 1990: Assisted/tutored for classes Acoustics (Fall 1988, 89), Signals and
Systems (Spring 1989, 90, and Fall 1989), and Circuits and Electronics (Spring 1990). Required
knowledge of course materials and lab equipment, and one-to-one contact with students.

Oak Ridge National Laboratory   Oak Ridge, TN
Engineering Physics and Mathematics Division
Summer 1989: Diagnosed and corrected problems in the vision routines related with accuracy
in determining the distance, angle, and yaw of an object from Hermies IIB, a robot running C,
68020 and 80286 Assembly languages on a UNIX based N-CUBE 4 parallel processor.

Summer 1988: Created a procedure that graphically represents the breakdown of a program for
execution on a parallel machine, along with a graphical interface between user and breakdown
programs, GAPPS (Graphing And Plotting on a Parallel System).

Summer 1987: Parallelized a medium- sized C program on a UNIX based N-CUBE 10 parallel
processor to estimate possible speedup of the serial code.

Papers   Einstein, J. R., C. W. Glover, and M. L. Dickens, "A Tool to Aid in Mapping Computational
Tasks to a Hypercube," CESAR Lab, ORNL. Appeared in the proceedings on the Fourth
Conference on Hypercubes, Concurrent Computers, and Applications, March 1989.

Beckerman, M., D. L. Barnett, M. L. Dickens, and C. R. Weisbin, "Robust Performance of
Multiple Tasks by a Mobile Robot," CESAR Lab, ORNL. Presented at the IEEE Conference
on Robotics and Automation, May 1990.

Interests   Letter in lightweight varsity crew. Sports include biking and swimming. Play bass and piano.

(b)

the earlier version (a), he inadvertently emphasized months and years rather than
achievements. Compare the two styles of referring to his work at Oak Ridge
National Laboratory. In the later version (b), he combines closely related activities and eliminates redundency.

Jeff David Sung
14988 El Soneto Drive
Whittier, CA 90605
email address: usungj@humanitas.ucsb.edu

June 7, 1997

Dr. John Krug
Vice President - Research and Development
Metech Laboratories
450 Technology Park Suite 44
Cambridge, MA 03687

Dear Dr. Krug:

I received my Ph.D. degree in Metallurgy from MIT earlier this month, and I am interested in career opportunities at Metech Laboratories. I have enclosed a copy of my resume, which will give you more information on my academic, research and industrial experience.

My general areas of interest include physical metallurgy and mechanical behavior of materials. My research experience has to date focused on fatigue and fracture. I am currently doing post-doctoral research at MIT on fatigue in 2024-T3. This work is sponsored by the Department of Transportation to address some of the fractographic issues that are relevant to the aging aircraft problem. My thesis work centered on the fatigue behavior of 6063-T5 extrusions for use in the vertical axis wind turbine test bed built by Sandia National Laboratory. This material was found to exhibit fracture by a ductile intergranular mechanism similar to that reported in Al-Li alloys.

The literature I have read points to some interesting research going on at Metech. I am also aware of the commitment to the evolution of the R&D process to improve your products and to your response to customer needs. Given my technical background and research experience, I believe I could make a contribution to Metech and learn a great deal at the same time.

I would welcome the opportunity to discuss employment opportunities at Metech Laboratories. I can be reached either at my office (310) 253-4744, or through my answering machine at home, (310) 926-7759. I look forward to hearing from you. Thank you for your consideration.

Sincerely,

Jeff David Sung

Jeff David Sung
Enclosure

Figure 15.2
The author of this letter of application to Metech Laboratories has skillfully taken the formula and shaped it to emphasize his qualifications and experience. The enclosure referred to is his resume, and the letter form is modified block. (Modified and used with permission, MIT Office of Career Services and Pre-professional Advising.)

large enough to accommodate your documents without folding and by designing a mailing label that matches your letterhead.

Reference and Follow-up Letters

Several additional documents may help you in your job search. If you plan to submit a list of references with an application, you should inform each person on the list of your plans and get explicit permission to use the person's name. A reference letter from someone who does not want to write it is not much good. You can get permission by telephone, of course. But we think that you will increase your chances of getting a letter that focuses on your fitness for the job if you write a letter to your referee, calling attention to aspects of your experience that match your potential employer's needs. Always enclose an updated resume. Inform your referees of the outcome of your job search and thank them for the time they have spent on your behalf.

Many job applicants write thank-you letters to companies where they have interviewed (Figure 15.3). Declining an offer presents another important occasion for a letter. If you decide to say no, a courteous acknowledgment is certainly in order (Figure 15.4). Writing to thank someone for an interview can, in tight circumstances, make the difference between being hired and being second in line. It suggests that you'll be considerate and pleasant to work with. Writing letters declining offers can also pay off, though sometimes years later. In the small networks that form any subspecialty, having the grace to say "no" formally may make you worth remembering when new projects come up.

The Future of CVs and Resumes

In the past few years, the resume form for technical professionals has changed in appearance, but the content has remained relatively constant. Because of the widespread availability of desktop publishing software and laser printing, resumes and CVs look much more attractive than they used to. If you have not updated the design of your resume, it may look underprepared in comparison with others.

Sarti Ahmed
410 Memorial Drive
Cambridge, MA 02139
email address: sarahmed@athens.mit.edu

September 10, 1994

36 Ames Street
Cambridge, MA 02139

Ms. Tina Oshima
Novaq Systems, Inc.
874 Wild Oak Ridge
Sacramento, Ca 95831

Dear Ms. Oshima:

It was a pleasure to have spoken with you and Mr. Mansfield yesterday, regarding job
opportunities at Nova Systems. The interview was informative, and I am very
interested in the work you are doing. My past two summer positions were related to
the development and design of software programs for industrial computer vision
experiments. I was very impressed with the latest advanced applications used in your
company. With my skills and interest in software design, I believe I could be of value
to Nova Systems.

Please let me know if you need any more information about my background. Thank
you for your time. I look foward to hearing from you.

Sincerely,

Sarti Ahmed

Figure 15.3
Letter expressing thanks for an interview.

Aparoopa Dutta
897 Main Street
Cambridge, MA 02139

May 16, 1994

23 Beacon Street
Boston, MA 02110

Mr. Timothy Richards
Director, College Relations
Northeast Electronics Laboratories
1 Water Works Plaza
Hudson, NY 34687

Dear Mr. Richards:

I am writing to thank you for the job offer to join Northeast Electronics Laboratories as a member of the research and development staff. Unfortunately, I must decline your offer. I have accepted a position with the energy-consulting firm of Charles River Conservation in Cambridge, MA.

It was a difficult decision for me because I was both excited and impressed by the work at Northeast Electronics. I appreciate your giving me the opportunity to meet with you and the members of the research staff.

Again, thank you for your time.

Sincerely,

Aparoopa Dutta

Aparoopa Dutta

Figure 15.4
Letter declining a job.

Several commercial services now disseminate electronic resumes, usually on-line versions of conventional resumes. With presently available technology, however, energetic applicants can design multimedia resumes to display on personal Web sites. These are typically hypercard stacks that begin with a conventional resume. Each element can be followed in depth, with amplification that includes text, sound, still picture, and moving picture. For example, the education designator might link to a list of all undergraduate courses you took, while the name of each course could link to a course description and syllabus. Possibilities are very broad: you could include full text of laboratory reports or other documents written in a course, or videos in which you demonstrate expertise by performing specific tasks.

Though very few employers expect a multimedia resume, it is easy to imagine a future in which the conventional resume will be obsolete. For now, however, an informative and well-designed resume or CV has many uses.

16

Document Design

For almost 5 years you've been drafting documentation for a start-up company. You've worked with all departments in compiling documents at each stage of product development. Your efforts have paid off. Now, as the company moves from development to marketing, you've been made the manager of a department, and you've just hired two new staff writers. Your first task is to instruct them about specifications for the company's documentation, most of which are stored only in your memory. You need quickly to develop document standards.

Technical documents are rarely built of words alone. In addition to sentences and paragraphs, they are likely to contain a mix of other elements: headings and subheadings, headers and footers, tables of contents, equations, appendixes, abbreviations and acronyms, figures, and tables. This mix creates design as well as writing tasks for *all* technical forms, including memos, letters, proposals, reports, procedures, transparencies, slides, and computer screens.

A high-quality document is *consistent* in the way elements are treated. Page numbers appear in the same location on every page. The same font is used for all first-level headings. Abbreviations, acronyms, and plural formations are standardized. All bulleted lists are indented the same number of spaces. Equations are either numbered or not numbered, but the practice is unchanged throughout all documents in a series.

Writers in government, corporate, and academic settings are rarely the first people in their organization to write memos or letters, to prepare a set of slides, or to bind progress reports. Many work settings already have standard formats for memos and letters, preprinted covers for reports, templates for transparencies and slides, and "house style" for settling questions about the treatment of oversized illustrations.

In some work settings, document standards are transmitted informally: authors ask colleagues who have already prepared similar documents, or they examine models of earlier work. In other settings, standards are transmitted through a written style guide. The style guide may have been especially prepared for authors in one organization. It may be a field-specific manual like the American Chemical Society's *Handbook for Authors,* or it may be a more general reference book like *The Chicago Manual of Style.* It may be a military specification (milspec) or a publication standard developed by the American National Standards Institute (ANSI).

Preparing a Style Guide

Whatever your writing task, begin with a clear idea of how finished pages will look and how the final document will be packaged. If you are working with coauthors, each member of the writing group must have the same instructions about the physical appearance of pages or screens. Individual authors will save time because they do not need to make style or format decisions. The resulting document will be consistent (Figure 16.1).

Thinking Visually

Visualize your completed document as a series of two-page spreads rather than a pile of individual sheets of paper. Two-page spreads are what readers will see when they read your document. Consider printing or photocopying on two sides so that readers receive facing pages of text and illustrations. Such a method not only looks more professional but also increases the possibilities for placing figures and tables on the same spread of pages in which they are discussed. Number prefatory pages with small roman numerals, pages in the report body with arabic numerals, and pages in the appendix with alphabetic designators (Figure 16.2).

Create a design for individual pages in the report body, and use it consistently. Will each page have a header? A footer? Will you print in one or two columns? How much space will you leave between paragraphs?

Headings, 12pt Bold Times | Center Justify

Subheads, 12pt Bold Times | Sub-subheads are tabbed 1 time (0.5" tabs), and each additional subhead is tabbed one more time

Main Text, 12pt Plain Times
• The document will be on 8.5" x 11" paper bound by a plastic cover, and each section will be tabbed.
• The document will be printed on facing pages.

Starting New Sections
• New sections will start on right facing pages.
• Left facing pages left blank will be marked "This page intentionally left blank."
• Paragraphs will not be indented.

Page Numbering
• Pages will be numbered on the bottom center of each page.
• Front matter will be numbered in lowercase roman numerals; the main body will use arabic numbers; the appendixes will use an uppercase letter and an arabic number separated by a dash (e.g. A-2).

Tables, Figures, Footnotes, and References
• Tables will be incorporated into the text; figures will be placed in an appendix.
• Footnotes will be located at the bottom of the page.
• References will be placed in the footnotes and numbered consecutively.
• All references will be numbered using arabic numbers in brackets.

Abbreviations, Acronyms, and Equations
• Any special abbreviations or terminology will be explained in the glossary.
• All acronyms will be spelled out in the glossary and also the first time they appear in the text.
• All equations will be written in mathematical notation; variables will be explained in the glossary.

Page Numbers
12pt Plain Times | Center Justify

Figure 16.1
This page format template creates standards for all documents produced at Cimarron Automation Services, Moorpark, Calif. Note that the style guide itself is prepared in the recommended format.

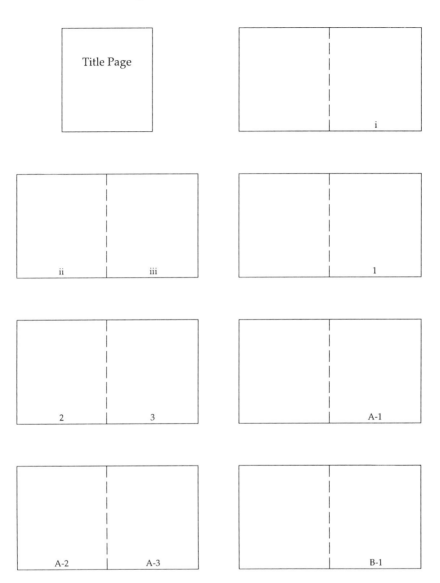

Figure 16.2
Plan documents as a series of two-page spreads, with expanded possibilities for placing illustrations and text on the same spread of pages. Note that prefatory, body, and appendix sections are clearly designated by the style of page numbers.

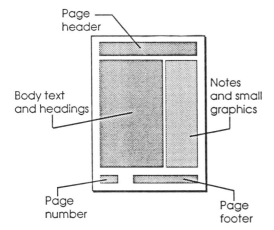

Figure 16.3
This page template would suit an instruction manual or a technical report with text, notes, and graphics. In this style, oversized illustrations are placed in an appendix.

Between subsections? Where will you place page numbers? Will the right-hand margin of text be justified or ragged (Figure 16.3).

Select a type style and size for each element in your report and use it consistently. Elements include titles, first- and second-level headings, text for the body of the report, legends and titles for figures and tables, and style for headers, footers, and references. Figure 16.4 illustrates a widely used style of headings.

Begin every major section of a report on a right-facing page, even if you need to leave extra space at the end of the preceding section. This strategy reinforces the modular organization of technical reports while providing readers with natural places to stop and restart.

Planning Illustrations

You need a repertoire of strategies to deal with tables and figures. Full-page illustrations that can be studied without turning the page are commonly called "portrait" figures; illustrations that are wider than they are tall have a "landscape" orientation, requiring readers to turn the page sideways. When an illustration requires a landscape presentation, place it so that it can be viewed by rotating the page clockwise.

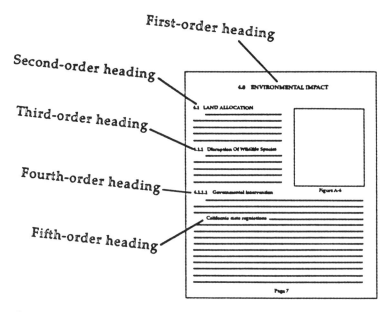

Figure 16.4
In heading style, upper- and lowercase typography, centering, indenting, and numbering systems serve as markers of the hierarchical levels of information.

A figure or table too large to be contained in a portrait or landscape page can be made into a foldout, an oversized page folded to fit the dimensions of the printed report (Figure 16.5). Foldouts can be prepared with or without an *apron*, a blank page that forms the part of the foldout nearest the report binding. Aprons allow readers to open the foldout and refer to the illustration while reading the text, rather than alternating back and forth.

Presenting the Product
Consider binding options. Comb and spiral bindings help readers to keep pages open and flat. Loose-leaf bindings are good choices for documents that must be updated with new pages. Heavyweight or plastic covers will help your document hold up through multiple readings. Tabbed section dividers help readers to navigate efficiently through your document. Innovative binding methods like the TRI-S-PORT allow for side-by-side display of text and appendix on a minimal surface (Figure 16.6).

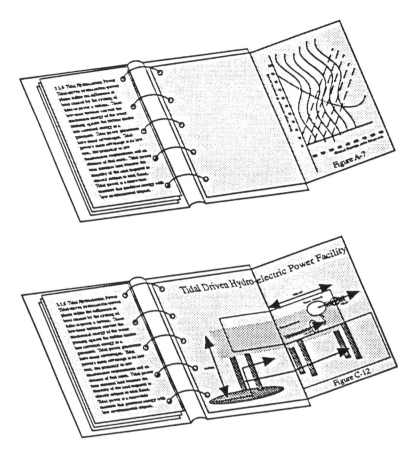

Figure 16.5
The top figure is prepared as a foldout with an apron. Readers can refer to the figure while they are examining other pages of the report. The bottom figure is a standard foldout of an oversized illustration.

Electronic Document Standards

The widespread use of computers to communicate scientific and technical information has created more interest than ever in standards for format and style. Required page formats can be stored as electronic templates. Writer's choices over matters of language, structure, and design elements can be limited so that consistency is achieved across an entire document or set of documents.

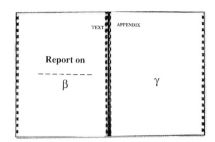

a. Closed version

b. Partly opened version

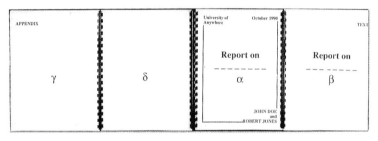

c. Fully opened version

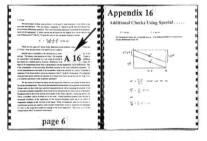

d. Text page 6, with Appendix 15, with "wings" folded under

e. Text page 6, with Appendix 16, with "wings" folded under

Figure 16.6
The TRI-S-PORT format for reports makes it possible to present text and illustrations on facing pages. (Used with permission of Dr. Stephen Juhasz.)

Document Templates

Many organizations achieve consistency of format, organization, style, and content by developing and using electronic document templates to accompany research and development projects. Writers are electronically constrained from violating standard formats. Though document templates do not eliminate all pains of authorship, they do answer such questions as what should be included, in what order, and at what level of detail. Document templates are frequently used in software development, where the need to document is urgent and sometimes overwhelming. A well-conceived set of templates provides outlines for the eventual full set of documents, from the early specification stage to the shipping of the product.

SGML and HTML

Standard General Markup Language (SGML) has enormous potential significance for enforcing standards in technical communication. Since 1986, SGML has been a standard of the International Organization for Standardization (ISO). It is the publishing system of choice for the U.S. Department of Defense, the American Association of Publishers, and the European Economic Community.

With SGML, information is independent of the program that created it. Text can be transferred from one computer system to another without loss of format. Electronic and print output can emerge from the same text base. A single document can be stored for later use and then combined with other media for mixed-media displays. Rules can be defined for document elements, and programs called "SGML parsers" then validate documents for consistent structural usage. The standard for SGML-compliant documents is defined in ISO 8879. A more recent ISO standard for hypermedia documents defines the HyperText Markup Language (HTML) used to write pages on the World Wide Web. Information about these new developments is available from the American National Standards Institute (http://www.ansi.org/) and ISO (http://www.iso.ch).

The Future of Document Standards

The term "document" now includes multimedia productions with sound and animation, proposals in book format, and pages on the World Wide

Web. A static document, designed, printed, and bound in a certain way, is only one of many formats in which the text can be presented. New publishing software separates content from format. Text can be updated quickly for print or electronic delivery. But achieving legible on-line text may take more effort than selecting a menu option that automatically converts a paper manual to a computer display. Despite advances in publication technology, authors still need to be aware of visual design features that facilitate reading.

17

Strategies for Searching the Literature

You've just begun a new job in the R&D department of a chemical products firm. You were hired to be part of a new team that will eventually develop a new product line. Before your project begins, however, you need to identify current research that affects your team's plan. You're equipped with electronic access to a variety of databases, but you have only a few weeks to search the literature and summarize how it applies to your project. Where do you begin?

Literature searching always involves a time trade-off. Locating published information can support your research and may even streamline parts of your work. Staying abreast of developments avoids duplication of findings, but the published record is gigantic. Spending time tracking down information may not be useful. You can waste valuable hours better spent talking with colleagues and supervisors and working on your project.

The Flow of Technical Information

Searching the published record helps you

- Locate reference information necessary for routine research
- Gather data to extend your methods, findings, and discussions
- Follow the broad trends in your field and identify promising research problems
- Follow the theoretical, methodological, and design work of related fields

Remember that information in print is cold data. It's not likely to be up to date. The same scientific knowledge is released in many forms over several years. More current information flows orally among colleagues in a laboratory, passing through small in-house seminars and sometimes becoming bottled in proposals, progress reports, and memos, most of which are proprietary. In industrial firms, the process often ends there, with limited circulation through a routing or mailing list. The information may also find its way into patents, specifications, manuals, and corporate bulletins.

In pure research, information moves outside the organization into various forums: conference proceedings, formal reports, and refereed articles. Moving from project initiation to journal publication takes up to 5 years. Four to 18 months may pass as the manuscript goes through the review-editorial stage and emerges as a published article. Three more months to a year may pass before the information is indexed and abstracted. Another 5 years may elapse before it is absorbed in reference works, including review articles and textbooks.

You can intercept information at several stages—in conversations, letters, seminars, colloquia, preliminary reports, theses, preprints, published reports, articles, literature guides, and reference works. Two factors vastly increase your ability to be current and concrete in your information searching:

• *Access to an expert.*   The closer you get to the source of the expertise, the more current the information.

• *Electronic data.*   On-line capabilities, both in local databases and on the World Wide Web, not only increase your reach, speed, and versatility in locating information but also give you access to information before it is in print.

The Reference Library

Libraries have three elements: catalogs, literature guides, and collections, or stacks (Figure 17.1). Catalogs, now mostly on-line, list the holdings of a library; they are your primary points of entry. While catalogs list journals and other serial publications, they do not normally list individual journal articles, reports, or other short forms. Short publications are

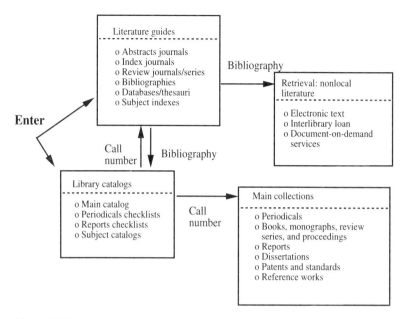

Figure 17.1
The library as a system.

indexed in literature guides for specific fields. No library, however, will contain every item listed in guides like *Chemical Abstracts* or *Engineering Index*.

Increasingly, catalogs, literature guides, and collections are incorporated in an electronically linked system. You enter the system from your workstation or personal computer, accessing either the on-line catalogs or the database of selected literature guides. From all these, you retrieve bibliographical entries and their call numbers, which you then use to locate items of interest, either in hard copy in the library's collections or through electronic access. The texts held in the library's main collections may be hard copy, microform, electronic files, or compact disk technology (CD-ROM). CD-ROM storage increases space, with a single disk capable of storing up to 300,000 pages of print. The current trend in information retrieval is toward the building of large-scale digital libraries that are accessed on the Internet.

```
          WELCOME TO BARTON, THE MIT LIBRARIES' ONLINE CATALOG

      Use arrow keys to highlight selection; then enter search terms.
         *OR* type command; then enter search terms:  t=moby dick


  Look for the FIRST WORD(S) in:              Look for KEYWORD(S) in:

      Author          a=                  Author Keyword    aw=
      Title           t=                  Title Keyword     tw=
      Subject         s=                  Subject Keyword   sw=
      All Indexes     all=                All Keywords      w=
      Call Number     c=                  Notes             nw=


                          <Series Search>
      <Additional Searches>                   <Boolean Searching>
      <Course Reserves>                       <Thesis Searching>

   >>>
        Enter author's name. (lastname, firstname)

           For HELP, highlight command and press <ENTER>.
              To EXIT, type E and press <ENTER>.
```

(a)

```
                                         Number of holdings :1
          TITLE :Lipids and tumors / ed. by K. K. Carroll ; contributors,
                 Francis M. Archibald ... [et al.].
       LANGUAGE :ENGLISH
      PUBLISHED :Basel ; New York : S. Karger, 1975.
  PHYSICAL DESC :x, 399 p. : ill. ; 25 cm.
         SERIES :Progress in biochemical pharmacology; v. 10
   BIBLIOGRAPHY :Includes bibliographical references and indexes.
        SUBJECT :Tumor lipids
                 Carcinogenesis
   OTHER AUTHOR :Carroll, Kenneth Kitchener.
                 Archibald, Francis M.
  LC CONTROL NO. :75318809
           ISBN :3805517084 : 148.00F


   LIBRARY          COLLECTION / CALL NUMBER          STATUS / DUE DATE
   -------------    -------------------------------------  ---------------
  1. SCIENCE        STACKS / RC268.5.L56                   In Library
  Last Page
  Options:    Go to previous screen/list
  Prior screen    Extend search    Brief display    Search copy    Output   MARC
  Order display   Review search    OPAC parms   New search    Reset    ? help
```

(b)

A typical on-line catalog entry for an author shows

• Author-title-subject information
• Publication and imprint information
• Call number

For example, the two on-line catalog screens shown in Figure 17.2 provide search options and search results for a volume titled *Lipids and Tumors*. On-line capabilities improve the speed and facility of library access. You can search through authors, titles, subjects, numbers (ISBNs, call numbers, government numbers, etc.), and keywords. Any of these categories can help you locate a desired text or help you concentrate sources of information.

Finding Technical Literature
You locate different kinds of documents by consulting general or specialized listings, including the main catalog and standard reference works. Consult reference librarians if you are unfamiliar with the guides that index the literature of your specialty.

Guides to the Literature Literature guides (e.g., *Physics Abstracts, Engineering Index*) list and abstract individual articles. These guides are indexed in the main catalog by title and corporate author (sponsoring organization). More than 2000 abstracts journals cover the annual research output of the sciences and applied sciences. These literature guides are listed in various reference works, including Chin-Chih Chen's *Scientific and Technical Information Sources* (1987), which arranges bibliographies and literature guides by field.

Abstracts journals cover mostly articles and reports but also include patents, theses, proceedings, and books. Some, like *Government Reports Announcements and Index*, are devoted to agency-sponsored research, usually issued as technical reports. Others, like *Computer and Control*

Figure 17.2
Typical on-line catalog entry. (a) Search selections. (b) Title search result. The searcher dials the search choices in (a) and carries out a title search for the book. The same volume could be found using any of the other several options listed in (a).

Abstracts or NASA *Scientific and Technical Aerospace Reports* (STAR), are devoted to a field with many subdisciplines. The *Science Citation Index* and *Index Medicus* are interdisciplinary in scope and often overlap. If your library subscribes to the electronic version of a literature guide like *Chemical Abstracts* or a database like *MEDLINE*, you can search it electronically—possibly from your office computer.

The *Engineering Index Annual* (EIA), a sample entry of which is shown in Figure 17.3, is an abstracts journal. The EIA abstracts are arranged by subject, in keeping with *Engineering Information Thesaurus*. The entry in Figure 17.3 appears under the main subject heading "Biomechanics," and the subheading "Joints," after which abstracts appear in numerical sequence. If you have access to COMPENDEX, the on-line version of this index, you can search for authors, titles, keywords, and institutions (Table 17.1). Each abstracts journal or service is arranged differently, with its format described in an introductory section.

Journals   Professional journals are usually listed in the main catalog, which identifies call numbers (e.g., Q.S399 for *Science Magazine*). Some libraries, especially those without on-line main catalogs, may list journals and other serial publications in a *Periodicals Checklist* located in a fiche file in the library's reference section. When you find an article listing in a literature guide, you then go to the main catalog or periodicals checklist to find the journal call number.

Books, Monographs, Proceedings, Review Series   Books of all kinds are listed in the main catalog under authors or editors, title, subjects, and corporate authors. Locating conference proceedings often requires the conference title and date. If you need these, consult the librarian. Proceedings are also listed in the Institute for Scientific Information's (ISI) *Guide to Conference Proceedings* and in other literature guides such as *Engineering Index*. Review series (e.g., *Advances in Bioengineering*) are listed under the series title).

Reports   Some reports are listed in the main catalog under author, corporate author, title, subject, or number. These entries, however, represent only a fraction of the report literature in a technical library. Normally,

Amvrosova, O.I., 102487
An, Chae, 073109
An, K.N., 011034
An, Quang Dieu, 044888
Anada, Tetsuo, 109026

(a) **Author index**

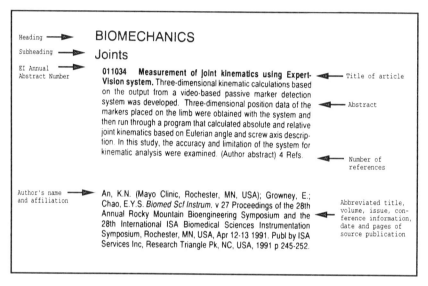

Heading ▶ **BIOMECHANICS**

Subheading ▶ Joints

EI Annual
Abstract Number ▶ **011034   Measurement of joint kinematics using Expert-Vision system.** Three-dimensional kinematic calculations based on the output from a video-based passive marker detection system was developed. Three-dimensional position data of the markers placed on the limb were obtained with the system and then run through a program that calculated absolute and relative joint kinematics based on Eulerian angle and screw axis description. In this study, the accuracy and limitation of the system for kinematic analysis were examined. (Author abstract) 4 Refs. ◀ Title of article

◀ Abstract

◀ Number of references

Author's name and affiliation ▶ An, K.N. (Mayo Clinic, Rochester, MN, USA); Growney, E.; Chao, E.Y.S. *Biomed Sci Instrum.* v 27 Proceedings of the 28th Annual Rocky Mountain Bioengineering Symposium and the 28th International ISA Biomedical Sciences Instrumentation Symposium, Rochester, MN, USA, Apr 12-13 1991. Publ by ISA Services Inc, Research Triangle Pk, NC, USA, 1991 p 245-252. ◀ Abbreviated title, volume, issue, conference information, date and pages of source publication

(b) **Sample entry**

Figure 17.3
Typical entry from an abstracts journal, the *Engineering Index Annual* (EIA). (a) Author index entry, (b) Main listing, a conference paper, which is part of an annual symposium series. The volume is the 27th of a proceedings series for the Biomedical Sciences Instrumentation Symposium. The searcher may also work with the EIA by subject heading, author, abstract number, keywords, or institutional affiliation. EIA abstracts are arranged by subject, in keeping with the *Engineering Index Thesaurus*. (Used with permission.)

Table 17.1
Some key information vendors and their services

Vendor	Publications	Database	Current awareness	World Wide Web: URL
BIOSIS (Biosciences Information Services) 2100 Arch St., Philadelphia, PA 19103	*Biological Abstracts BA/Reports, Reviews, Meetings Zoological Record*	BIOSIS TOXLINE	✓	http://www. biosis.org/
CAS (Chemical Abstracts Service) 2540 Olentangy River Road, Columbus, OH 43210	*Chemical Abstracts Parent Compound Handbook Registry Handbook International Coden Directory CAS Source Index*	*STN International Chemcats Chemlist CA plus online CAS Bio Tech Updates Chemical Patents Plus*	✓	http://info. cas.org/
DTIC (Defense Technical Information Center), 8725 John J. Kingman Road, Suite 0944, Fort Belvoir, VA 22060	*Technical Abstract Bulletin (TAB) DTIC Digest*	DROLS (Defense RDT&E online system) STINET	✓	http://www. dtic.dla.mil/
Engineering Information, Inc, Castle Point on the Hudson, Hoboken, NJ 07030	*Engineering Index Energy Abstracts Bioengineering Abstracts*	Ei COMPENDEX (computerized engineering index)	✓	http://www. ei.org/
INSPEC (Information Services in Physics, Electrotechnology, Computers, and Control), IEE, Savoy Place, London WC2R OBL, UK	*Science Abstracts: Series A: Physics Abstracts Series B: Electrical and Electronics Abstracts Series C: Computer and Control Abstracts*	INSPEC	✓	http://www. iee. org.uk/
ISI (Institute for Scientific Information, 3501 Market St., Philadelphia, PA 19104	*Science Citation Index Index to Scientific and technical Proceedings Index to Scientific Book Contents Index to Book Reviews in the Sciences*	Scisearch *Current Contents*	✓	http://www. isinet.com/
MEDLARS (Medical Literature Analysis and Retrieval System), National Library of Medicine, 8600 Rockville Pike, Bethesda, MD 20894	*Index Medicus NLM Current Catalog*	MEDLINE (Medlars online) TOXLINE (toxicology Information online) CANCERLIT, EPIPLEPSYLINE, etc.	✓	http://www. nlm.nih.gov/
NTIS (National Technical Information Service), U.S. Dept. of Commerce, 5285 Port Royal Road, Springfield VA 22161	*Government Reports Announcements and Index Government Inventions for Licensing etc.*	NTISearch *Federal Research in Progress*	✓	http://www. fedworld.gov/ ntis/

libraries maintain a separate reports checklist, either on hard copy or on-line, alphanumerically arranged by report number or government number. Once you locate a report and its number in a literature guide, you identify its location by finding the report number in your library's reports checklist. For example, the BNL series of Brookhaven National Laboratory is listed before the BTL series of Bell Telephone Laboratories, which in turn precedes the NASA series of the National Aeronautics and Space Administration.

Dissertations   In academic libraries, dissertations written at the same institution are indexed in the main catalog under author and title. Other dissertations in science and applied science appear in the *Dissertation Abstracts International, B, The Sciences and Engineering*. This reference work is usually in a fiche file in the reference section of your library and is also available on CD-ROM.

Standards and Patents   The vast literature of standards and patents is too diffusely distributed for all but a highly specialized library. Indexes to the patent and standards literature are often listed in the main catalog under subject headings like "Patents" and "Standards." Consult the reference librarian.

Electronic Journals, Bulletins, and Discussion Lists   Electronic journals and discussion lists are proliferating. Some are useful; others are not. Electronic networks referee and disseminate research results, but electronic publication in the sciences and applied sciences is still limited by lack of easily retrieved sophisticated graphics. Network bulletin boards and discussion lists do provide access to ongoing technical discussion, conference announcements, job lists, news in the profession, software, language groups, and so on.

Planning Your Search

Your search strategy depends on your task. Looking up a handbook to locate a physical constant differs from the months-long processes of reading, note taking, and building a comprehensive bibliography in a specific area. Always plan your search before you commit a lot of time to

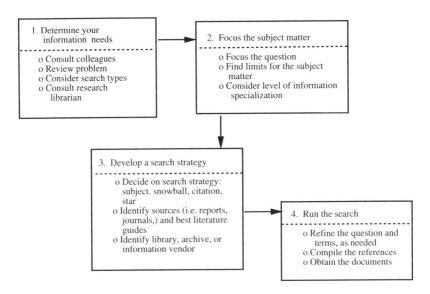

Figure 17.4
Steps in a search process.

the process. If the search is not routine, you can save time and avoid becoming bogged down by consulting a research librarian. View literature searching in stages (Figure 17.4).

Determining Your Information Needs

Ask yourself what aspects of your problem might be explained in the published record. Think carefully about what you want to accomplish. You can spend hours flipping through catalog cards or wandering around electronic databases, hoping to find a useful listing. Much— maybe most—information will come to you from colleagues and by word of mouth. You can add to this information by focusing your search on a specific goal, like one of the following:

- *Bibliography search.* To fill out a citation or check its accuracy
- *Location search.* To retrieve a published item
- *Subject or concept search.* To isolate a class of information by using subject headings and keywords
- *Methodology search.* To find information about processes invented and refined by others

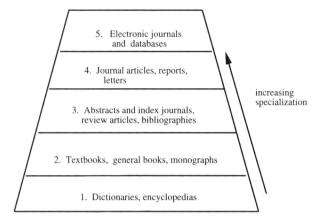

Figure 17.5
Increasing levels of specialization in studying a new field. If you are new to a subject area, you would normally consult the broader references and then work upward. Note that more specialized forms also tend to be more recent.

- *Follow-up search.*   To trace developments in the theory, applications, or results of a field
- *Specific question search.*   To find an answer to a specific question
- *State-of-the-art search.*   To identify the most recent advances in theory or applications for a specific concept or process
- *Multidisciplinary search.*   To concentrate information from sources across disparate fields
- *Comprehensive bibliography search.*   To compile an exhaustive list of sources treating a specific topic
- *World Wide Web search.*   Information search on the Internet using a web searcher such as *Yahoo, Lycos, Alta Vista*, or *Magellan*.

A search can produce a single item on a computer screen or a massive listing of hundreds of items. If you are new to a field, you may need to read background literature so that you can talk intelligently about a topic. Start with broad sources and progress toward more specialized works (Figure 17.5).

Focusing the Subject Matter
Your search needs a focused question. Often, you can make a broad question more useful by rephrasing it. For example, the question "How good is the available underwater connector technology?" will produce

more useful references if you rephrase it to ask "What was published on performance and reliability for underwater electrical connectors in 1982–92?" The word "good" is now expressed as "performance" and "reliability," two key terms widely used in the field. "Underwater connector" is qualified by "electrical." The rephrased question also establishes time limits.

The less abstract and open-ended the question, the more likely you'll get concrete references. You can thus focus a search by refining your terms. If you are uncertain of the key terms for a topic, consult one of the three standard thesauri for science and engineering: *Thesaurus of Engineering and Scientific Terms* (1967), *Library of Congress Subject Headings* (1989), or the *Engineering Information Thesaurus* (Milstead, 1992) of the *Engineering Index*. Most keywords and subject headings in on-line catalogs and databases are based on one of these three guides.

Limit your search by considering the following:

- *Subject.* Key terms and subject headings
- *Sources.* Key publications, literature guides, or series
- *Time.* Inclusive dates for acceptable publications
- *Authors.* Specific authors or corporate sponsors of interest
- *Institutions.* Publications by individuals at key institutions
- *Documents.* Specific kinds of documents (e.g., patents, standards, reports)

You need not invoke all these limits in one search, but many literature guides and databases provide indexes and options that treat them all.

Developing Your Search Strategy

Searching for a specific publication is much easier than searching for general information. A specific publication requires you to locate a printed object. A general search requires you to concentrate information. When you have determined what you are looking for, you may decide just to glance at the indexes of a few relevant journals. You may also decide to pursue a more systematic strategy through literature guides and databases.

Ask colleagues and supervisors about literature guides and types of publications. Also check with reference librarians. For example, if research reports from the Naval Research Laboratory in Orlando, Florida,

```
TITLE:     IEEE transactions on medical imaging.
IMPRINT:   New York, NY : Institute of Electrical and Electronics Engineers,
           c1982-
PHYSICAL FEATURES: v. : ill. ; 28 cm.
NOTES:     Quarterly. * Indexed by Electronics and communications abstracts
           journal (Riverdale) 0361-3313 * ISMEC bulletin 0306-0039 *
           Pollution abstracts with indexes 0032-3624 * Safety science
           abstracts journal 0160-1342 * Index to IEEE publications 0099-1368
           * Life sciences collection * Computer & control abstracts 0036-8113
           July 1982- * Electrical & electronics abstracts 0036-8105 July
           1982- * Physics abstracts. Science abstracts. Series A 0036-8091
           July 1982- * Sponsored by the IEEE Acoustics, Speech, and Signal
           Processing Society ... [et al.]. * Title from cover.
OTHER AUTHORS, ETC:  Institute of Electrical and Electronics Engineers. *
           IEEE Acoustics, Speech, and Signal Processing Society.
OTHER TITLES:  Abbreviated title:IEEE trans. med. imag.
 LC CARD:  83-640807
 ISBN:     ISSN:0278-0062
```

Figure 17.6
Online catalog entry for a quarterly journal title, *IEEE Transactions on Medical Imaging*. At least 10 different abstracts journals cover the articles published in this journal.

are possible sources, the *Government Reports Announcements and Index* will be a useful literature guide. If articles in the journal *IEEE Transactions on Medical Imaging* are important to your work, you might want to see what abstracts journals index its articles.

Information about where specific journals are indexed is also provided in *Ulrich's International Periodicals Directory* located in the reference section of your library. *Ulrich's* is available in both hard copy and CD-ROM. As Figure 17.6 shows, *IEEE Transactions on Medical Imaging* is indexed in several different places. To trace a specific author, look in the author index of a literature guide. Electronic searching is versatile. Your query goes directly to the electronic files.

Subject Searching Subject searching means using subject headings or keywords to trace documents. With keywords, you can search either titles or subject areas. You identify keywords in books, articles, or thesauri. You then search the database, on-line catalog, or card catalog. Subject searches can be useful when you don't know much about the subject or when you just want to browse. Subject searching is also an excellent cross-disciplinary approach because the cross-references often show topical relationships between materials you don't normally associate.

In the sciences and applied sciences, however, the number of terms is so vast and expands at such a rapid rate that you need to use thesauri if you want to be accurate. In spite of its comfortable, encyclopedia-like feel, subject searching is often not the best technique. It's slow, even on-line, and it often produces barren lists of documents with little relevance to your interests. Subject searching on a Web searcher can produce thousands of listings.

Sometimes, you can address this weakness by using review articles, reference guides organized by subject. Review articles identify and summarize key articles and other publications that have contributed to the development of a specific field. Some review articles are broad surveys of several years' work. Others are state-of-the-art accounts of annual or semiannual research. Review articles appear in standard refereed journals, in special review journals, or in bound volumes with titles such as *Advances in* _____ or *Progress in* _____. Their unusually large numbers of references can help you identify review articles in literature guides.

Snowball Searching   The most widely used searching technique, the "snowball approach," begins with a recent publication. You find a key paper, preprint, review article, or textbook. Then you look up items listed in the bibliography. From those retrieved items, you look up further entries, and so on, as Figure 17.7a shows.

The snowball search is fast and requires little use of literature guides. You assemble a list of references that supply you with more references. This technique does, however, have limitations. It tends to move you back to literature that is obsolete, and if you begin with a marginal article, you can spend much time assembling a network of similarly marginal papers.

Citation Searching   In a citation search, you begin with a key source paper and compile a list of papers citing that paper. The basis of your search is that papers citing the source will be related. Just as the snowball

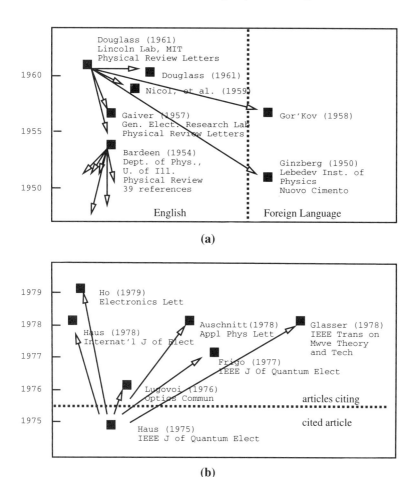

Figure 17.7
Two reference searches. The snowball search (a) moves the search back in time. An article by Douglass in *Physical Review Letters* (1961) takes you back to Nicol (1959), Gaiver (1957), and Ginzberg (1950). The article by Gaiver, also published in *Physical Review Letters*, takes you back to a review article with 39 references, published by Bardeen and in *Physical Review* (1954). In this way, you quickly find background literature and build a network of references. The citation search in (b) moves you forward. Six different articles cite a source paper by Haus, which appeared in *IEEE Journal of Quantum Electronics* in 1975. The most recent paper, by Ho in *Electronics Letters*, brings you forward to 1979.

search moves you backward, the citation search brings you forward because the papers citing are more recent than the paper cited, as Figure 17.7b shows.

The crucial literature guide for citation searching is the *Science Citation Index* (SCI) of the ISI. It is available in hard copy, on-line, and on CD-ROM. The SCI consists of a citation index that lists the authors cited in footnotes and bibliographies of selected journals and books and a source index that lists authors of all citing publications. Covering a core of more than 2000 journals, the SCI manages to account for a considerable percentage of the articles cited in basic science and applied science.

Citation searching offers distinct advantages. Like the snowball method, the citation search moves directly from document to document, with no intervening terminology or subject indexes. But the journals it covers are only a fraction of the published literature. It also has only limited applications to engineering.

"Star" Searching   With the "star" approach to searching, you look at certain star journals, researchers, or institutions that often account for a high proportion of the important publications and know-how in a specialized area. If you monitor their output, you will find critical work done in a given field. This approach offers focus.

To follow key journals, go to back issues, skim the article titles and abstracts, and glance over the cumulative subject indexes. To follow key authors, turn to the cumulative author indexes of journals or literature guides. You can also follow the research output of a facility known for its special work. For example, if a type of polymer research is being carried out in the materials science department at Case Western Reserve University, you can track that research in the corporate index of *Chemical Abstracts*. Corporate indexes are standard features of many literature guides. Key authors and institutions may be searched together.

Comprehensive Database Searching   You can carry out comprehensive database searches from a computer linked to one or more commercial database systems such as those listed in Table 17.1. You focus the search by using thesauri to develop a search profile of terms that will draw the desired information from the database. *On-line Database Selection*

(1989) lists several hundred databases that cover many technical fields and special forms of literature such as reports, theses, and patents. Using these databases can be intimidating and time consuming. The operation usually requires the assistance of a reference librarian.

Normally, the initial search profile needs to be tried out and focused until relevant titles and abstracts show on the screen. Add terms to narrow the search. Delete terms to broaden it. Database searching is useful when you need to to compile a bibliography, gain access to restricted databases, or find recent bibliographical information before it appears in literature guides. The sheer time required to read and assess the results of this kind of search, however, can make the process prohibitive.

Running the Search

Searches rarely go according to plan, so you need to combine planning with initiative. You might begin, for example, with one or two papers furnished by a colleague, retrieve selected references listed in those papers, and then refer to a literature guide that indexes journals in which the most interesting papers appear. The key terms in the literature guide may help you modify your own list of keywords. If you find a valuable author, you might look in the literature guide's author index to see what else that author has published. Examine the corporate index to see if the author's research group is producing other papers on the subject. You might also glance at the SCI to see what other publications have cited your key papers.

When you have compiled a list of potential papers, you need to retrieve them. This process can be time consuming. Examine the abstracts of the papers carefully to see if you really need the complete work. Papers on topics close to your own research may be especially important if you're in danger of duplicating the research of others. Photocopy and collect important papers. Note other references of limited interest. Keep your notes up to date.

Current Awareness Searching

Because advances and innovation are the lifeblood of scientific professions, staying informed is critical. The best way is through routine, informal oral exchanges with colleagues, as well as peer group discussions at professional and trade conferences. Formal current awareness searching, by contrast, can keep you abreast of developments outside your circle of acquaintances. It also requires special literature guides.

One simple but useful tool is the contents journal, which prints the contents pages of recent professional journals. *Current Contents*, a series of small biweekly contents journals published by the ISI, comes in different specialized fields. One is the *Physical and Chemical Sciences Edition*. Another, the biweekly publication of Chemical Abstracts Service, called *Chemical Titles*, covers about 700 journals. All the information vendors listed in Table 17.1 offer extensive current awareness services, which are described at their respective Web site URL's.

A related service is the published search. The National Technical Information Service (NTIS) annually publishes thousands of prepared searches, each based on a set of keywords that identifies specific subjects of interest. These bibliographies, derived from the NTIS database, may be ordered from the NTIS annual catalog, *Current Published Searches*. The ISI also provides *Ascatopics,* a series of bibliographical reports in more than 300 subject areas (see Table 17.1).

Information Vendors

Information is now a commodity, and big information vendors offer a variety of services. The most specialized (and expensive) are the Selective Dissemination of Information (SDI) services, which tailor searches to the individual purchaser. SDI services, offered by most information vendors, periodically match a database to search keywords provided by the purchaser. The results are mailed or transmitted electronically. These services are also available for new patents, government standards, and military specifications.

Available services include

• Published reference works such as abstracts and index journals, bibliographies, and other compilations
• Current-awareness services, including contents journals and SDI services

• General and specialized databases available for on-line searching
• Document-on-demand acquisition services by e-mail, telephone and by mail

Services available from nine information vendors are listed in Table 17.1. Each vendor supplies literature and maintains a Web site that outlines specific services.

18

Citation and Reference Styles

For the proposal you're about to finish, you have just agreed to write the literature review. Your job will be to summarize the research that has led the principal investigator to frame the question and apply for research support. You know the science involved, and you're familiar with 10 years of technical literature. The content should be easy. But now that you begin to write, you wonder how much to quote, where to cite a publication, or whether much of the summary is so basic that it needs no references.

Proposals, reports, and journal articles almost always contain references to earlier published work. Your work acquires credibility when you review the literature and show that your contribution extends from a solid foundation of respected research. When you reveal your sources, you enable readers to locate and consult the papers that informed your work.

If you have done a good job of researching your subject, the quality and quantity of the sources you have consulted will enhance your work. In citing sources and preparing an accurate reference section, you make it possible for readers to retrace your steps. They may wish to assess your work in the light of previous contributions to the subject, or they may be reading for ideas to help them with new or continuing projects. Reference sections often interest readers long after the research methods reported in the article have been superseded by new approaches. A good reference section provides access to the history of a technical problem. Your references can be as valuable as your research methods and findings.

In reporting the results of a literature review, you need to write citations to the earlier work you are reporting and to prepare a reference section or bibliography with specific details about authorship, publication, editions, and dates for each citation. At least three citation styles and 100 reference styles are commonly used. The appropriateness of each is determined solely by the expectations of potential readers. Use whatever style you choose consistently throughout a document, but know that the particulars of style are established by convention and group practice.

Find out if style guidelines are available in your work setting or provided by your anticipated audience. If you are submitting an article for journal publication, check the journal for advice about preferred reference style. Consider purchasing the style manual of your professional society. The American Chemical Society, American Mathematical Society, American Medical Association, American National Standards Institute, American Institute of Physics, and the Council of Biology Editors are some of the groups that publish guidelines for authors. Many of these organizations provide extensive information on World Wide Web sites.

Several software packages are available for managing references. You create a database for each item, storing bibliographic information such as author, title, date, publisher, place of publication, volume, number, and page. When you prepare in-text citations and reference sections, you can choose from a menu of styles, and the software will automatically format your paper. If you revise your paper for a publication with different documentation standards, the software will reformat as necessary.

In-Text Citations

In scientific and engineering writing, three styles are most commonly used for referring to earlier work: author/date (Figure 18.1); numbered references (Figure 18.2); and footnote or endnote citations (Figure 18.3). In each case, the citation provides a cross-reference to more complete bibliographical information provided at the end of the document in the reference section.

In selecting a style for your in-text citations, consider these comparative features. Many readers prefer author/date citations in a literature review. When names and dates are embedded in the text, readers immediately know what researchers you have consulted and can assess

The contribution of morphometric studies to octopus biology, and that of other soft-bodied organisms, is unknown because workers have traditionally assumed that preservation deforms octopus morphology (Robson, 1929; Vecchione, Roper & Sweeney, 1989). Demonstration that preservation-linked deformation of specimens is minimal (Voight, 1991a) opens this character set to new investigations. I compare external morphology among shallow-water species....

REFERENCES

Robson, G. C. (1929). *A monograph of the Recent Cephalopoda. Part I. The Octopodinae.* London: Brit. Mus. (Nat. Hist.).

Vecchione, M., Roper, C. F. E. & Sweeney, M. J. (1989). Marine flora and fauna of the eastern United States. Mollusca: Cephalopoda. *NOAA Tech. Rep.* NMFS No. 73: 1-23.

Voight, J. R. (1988a). Trans-Panamanian geminate octopods (Mollusca: Cephalopoda). *Malacologia* **29**: 289-294.

Voight, J. R. (1988b). A technique for trapping sandflat octopuses. *Am. Maloc. Bull.* **6**: 45-48.

Voight, J. R. (1991a). Morphological variation in octopod specmens: Reassessing the assumption of preservation-induced deformation. *Malacologia* **33**: 241-253.

Voight, J. R. (1991b). Enlarged suckers as an indicator of male maturity in *Octopus. Bull. mar. Sci.* **49**: 98-106.

Figure 18.1
Author/date citation style. In the author/date style, author name(s) and year of publication appear in parentheses in the body of the paper. References are listed alphabetically at the end, by authors' last names. If the author has published more than one item in a calendar year, each item is distinguished by a lowercase letter. (Source: Voight 1994)

the currentness of your search. Author/date citations, with alphabetical reference lists, allow easy additions to copy that needs updating before publication. Numbered references do not intrude on the text, so they are easier to skip over if you are reading strictly for research findings. If they are handled sequentially, though, numbered references may present problems when you add or delete anything. Footnotes and endnotes allow for additional, subordinated explanation that the writer might want to include, but not with the same prominence of the text.

The common modalities are surgery, chemotherapy, and X-ray therapy [1]. A particular modality used alone or in conjunction with another modality is hyperthermia [2, 4], in which a tumor is heated to diminish it. For localized tumor heating, multiple applicator phased arrays are commonly used [3] to transmit a focused radio frequency main beam at the tumor site [4, 5]....

REFERENCES

1. C.A. Perez and L.W. Brady, *Principles and Practice of Radiation Oncology* (J.B. Lippincott, Philadelphia, 1987).
2. S.B. Field and J.W. Hand, *An Introduction to the Practical Aspects of Clinical Hyperthermia* (Taylor & Francis, New York, 1990).
3. A.W. Guy, "History of Biological Effects andMedical Applications of Microwave Energy," *IEEE Trans. Microwave Theory Tech.* **32**, 1182 (1984).
4. J. Overgaard, "Clinical Hyperthermia---An Update," *Radiation Research, Proc. 8th Int. Congress of Radiation Research* **2**, *Jul. 1987*, eds. E.M. Fielden, J.F. Fowler, J.H. Henddry, and D. Scott, p. 942.
5. J. Overgaard, "The Effect of Local Hyperthermia Alone, and in Combination with Radiation on Solid Tumors," in *Cancer Therapy by Hyperthermia and Radiation, Proc. 2nd Int. Symp., 2--4 June 1977*, ed. C. Streffer (Urban & Schwarzenberg, Baltimore, 1978), p. 49.

Figure 18.2
Citations with a numbered reference list. In the numbered reference list style of citation, numbers are enclosed in square brackets rather than parentheses, to distinguish citations from equations. Each number refers to an item in the final reference list. (Source: Fenn et al. 1992)

Preparing a Reference List

The precise bibliographic form for items in your reference list will be established by the style guide you select. For author/date citations, references appear in alphabetical order. For numbered citations, references usually appear in the order in which you have referred to them, though some journals prefer alphabetical order in these cases as well.

If you have no other instructions to follow, here are some typical bibliographic forms:

Nereistoxin, a poison from the marine worm *Lumbriconereis heteropoda* is one of the earlier examples of this type of use and led to the synthesis of the insecticide cartap.[1] Thiocyclam, another insecticide, is a closely related structure. Better known and certainly more important are the pyrethrins, extracted from the flowers of *Chrysanthemum cinerariaefolium* Vis., which gave rise to the pyrethroid class of insecticides.[2] However, it is particularly pleasing to see two newer examples in the literature, as described below. The initial reports of the fungicidal activity of the strobilurins and oudemansins have led to a very large amount of interest.[3,4] Over 100 primary patents have now been filed, with some twelve companies involved, and the prospects for the commercial exploitation of this type of chemistry appear to be good in view of the reports at the recent British Crop Protection Conference.[5,6]

ENDNOTES

1. Eldefrawi, A. I., In *Insecticide Biochemistry and Physiology*, ed. C. F. Wilkinson. Plenum Press, New York, 1976, pp. 297-326.
2. Elliott, M., *Pestic. Sci.*, **27** (1989) 337-51.
3. Anke, T. & Steglich, W., In *Biologically Active Molecules*, ed. U. P. Schlunegger. Springer-Verlag, Berlin, 1989, pp. 1-25.
4. Beautement, K., Clough, J. M., de Fraine, P. J. & Godfrey, C. R. A., *Pestic. Sci.*, **31** (1991) 499-519.
5. Ammermann, E., Lorenz, B., Schelberger, K., Wenderoth, B., Sauter, H. & Rentezea, C., *Brighton Crop Protect. Conf., Pests and Diseases—1992*, Vol. 1. British Crop Protection Council, Farnham, UK, 1992, pp. 403-10.
6. Godwin, J. R., Anthony, V. M., Clough, J. M. & Godfrey, C. R. A., *Brighton Crop Protect. Conf., Pests and Diseases—1992*, Vol. 1. British Crop Protection Council, Farnham, UK, 1992, pp. 435-42.

Figure 18.3
Footnotes or endnotes. In the footnote or endnote style, a superscript number in the text corresponds to information provided at the foot of the page or the end of the chapter, section, or document. (Source: Pillmoor et al. 1993)

Book   McClintock, F. A., and A. S. Argon. 1966. *The Mechanical Behavior of Materials*. Reading, MA: Addison-Wesley.

Journal article   Fenn, A. J., and G. A. King. 1992. Adaptive Nulling in the Hyperthermia Treatment of Cancer. Lincoln Lab Journal 5:223–240.

Article in an edited collection   Engelbart, D. C. 1960. A Conceptual Framework for the Augmentation of Man's Intellect. In *Computer-Supported Cooperative Work*, edited by I. Greif. San Mateo, CA: Morgan Kaufmann Publishers.

Report   Benson, C. R. 1997. The Use of Single-Page, Direct-Manipulation Interfaces in Real-Time Supervisory Control Systems. Report No. 89-3. Santa Barbara, CA: Infotek Corporation.

Dissertation/thesis   Rariton, J. L. 1975. Parametric Instabilities and ElectrostaticIon-Cyclotron Waves in Multispecies Plasmas. Ph.D. dissertation, Massachusetts Institute of Technology.

Private Communication   Wachman, H. 1995. Private communication. Cambridge, MA: MIT Department of Aeronautics and Astronautics.

Conference paper in proceedings   Fettwels, A., and M. Nossek. 1981. Sampling Rate Increase and Decrease in Wave Digital Filters. *Proceedings, 6th IEEE Symposium on Circuits and Systems*, Chicago. 3:839–841.

Unpublished conference paper   Nathanson, A. 1998. New Tools for Engineering Artwork. Paper presented at the Annual Conference of the Society for Technical Publishing, Santa Rosita, CA.

Patent   Edgerton, H. E. August 16, 1949. Electric System. U.S. Patent 2,478,901.

Standard   American Society for Testing and Materials (ASTM). 1979. Standard for Metric Practice. PCN 06-503807-4l.

Citing Sources for Tables and Figures

If you photocopy, scan, or otherwise reproduce a figure or table from another publication for inclusion in your document, you must credit the source. Even if you redraw the illustration but borrow significantly from the original, you need to cite the original author and publication. References to figures and tables can be identified by the word *source* (often in small capital letters) and treated as an integral part of the artwork (Figure 18.4). For another option, source notes can be provided in the figure legend or table title.

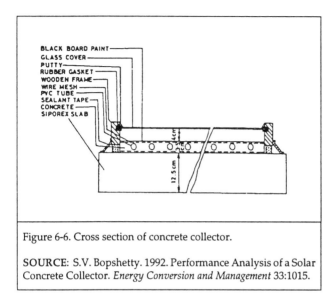

Figure 6-6. Cross section of concrete collector.

SOURCE: S.V. Bopshetty. 1992. Performance Analysis of a Solar Concrete Collector. *Energy Conversion and Management* 33:1015.

Figure 18.4
When tables and figures are reproduced from their original sources, the source is identified directly under the title or legend at the end. In most cases, the original needs to be rewritten for its new context.

Quoting and Paraphrasing Ideas from Sources

Even in documents explicitly designated as literature reviews, technical authors do not usually quote extensively from their sources. Rather, they summarize key points, restating ideas in their own words. Whether you quote exactly and enclose the borrowed phrases or sentences in quotation marks or you restate and summarize in your own language, you must credit the source. Only when the ideas and information are considered common knowledge should you avoid a citation. A simple test of whether ideas are common knowledge is this one: would this idea or piece of information be familiar to someone with your academic and professional standing (perhaps a colleague) who has not researched the subject? If the answer is yes, you do not have to cite the source. Otherwise, you must indicate the source of the material, even if it appears in several texts.

Short Direct Quotations

If a quotation occupies four lines or fewer in your manuscript, incorporate it in your text and use quotation marks to indicate where your words stop and the quotation begins. In the author/date format, provide a page number within the parentheses:

Both electric-field nulling, which reduces the "occurrence of undesired hot spots inside or on the surface of a target body," and electric-field focusing are "intended to maximize the RF power delivered to a tumor site" (Fenn and King 1992, 236).

Note that here the page number is not preceded by the word *page* or by *p*.

Long Direct Quotations

If a quotation is five lines or longer, set it off from the text by beginning a new line and indenting 1 inch from the left margin. Do not use quotation marks. If you are using author/date citations, do provide an exact page reference (Figure 18.5).

Partial Direct Quotations

Use ellipsis marks, usually three spaced periods, to indicate any deletions you have made in the quotation:

Fenn and King have demonstrated "electric-field nulling ... to reduce the occurrence of undesired hot spots inside or on the surface of a target body ... and electric-field focusing ... to maximize the RF power delivered to a tumor site" (1992, 236).

Altered Quotations

Use square brackets to mark any alterations you have made to the quotation. In the following example, the author is quoting directly from Fenn and King's paper, but she has added to the original source an explanation of the abbreviation RF:

Fenn and King have demonstrated "electric-field nulling ... to reduce the occurrence of undesired hot spots inside or on the surface of a target body ... and electric-field focusing ... to maximize the RF [radio frequency] power delivered to a tumor site" (1992, 236).

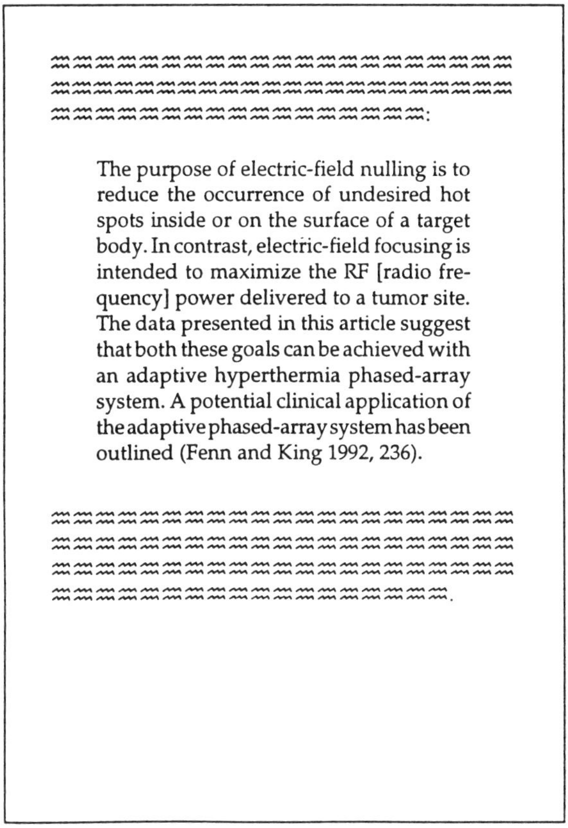

~~~~~~
~~~~~~
~~~~~~~~~~~~~~~~~~~~~~~~~~~~~~~:

> The purpose of electric-field nulling is to
> reduce the occurrence of undesired hot
> spots inside or on the surface of a target
> body. In contrast, electric-field focusing is
> intended to maximize the RF [radio fre-
> quency] power delivered to a tumor site.
> The data presented in this article suggest
> that both these goals can be achieved with
> an adaptive hyperthermia phased-array
> system. A potential clinical application of
> the adaptive phased-array system has been
> outlined (Fenn and King 1992, 236).

~~~~~~
~~~~~~
~~~~~~
~~~~~~~~~~~~~~~~~~~~~~~~~~~~~~~~~~~~~.

Figure 18.5
Long direct quotations are set off from the text. (Source: Fenn and King 1992)

Paraphrases

When you paraphrase, you restate ideas in new forms that are original in both sentence structure and word choice. Taking the basic structure from a source and substituting a few words is not an acceptable paraphrase and may be construed as plagiarism. Similarly, creating a new sentence by merging the wording of two or more sources is also unacceptable.

For the long direct quotation presented in Figure 18.5, the following restatement is not acceptable as paraphrase because it is too close to the original:

Electric-field nulling reduces unwanted hot spots in or on the target body; electric-field focusing maximizes the delivered RF power. These goals can be achieved by using an adaptive hyperthermia phased-array system (Fenn and King 1992, 236).

Here is an acceptable paraphrase:

During heating of a tumor site, an adaptive hyperthermia phased-array system can achieve two important clinical goals: reduced incidence of unwanted hot spots and improved power delivery (Fenn and King 1992, 236).

Note that both the sentence structure and some of the wording have been changed, so that quotation marks are unnecessary.

Citing Electronic Information Sources

Bibliographic forms for information gleaned from electronic media, including on-line journals and e-mail, are "under construction," but the purpose of any reference section is still to enable readers to locate and examine all documents referred to by the authors. How will readers acquire access to documents that exist as computer files? Technology has thus far advanced more rapidly than bibliographic practice, but some reference conventions are emerging.

In citing an electronic publication, include the publication medium (on-line database or CD-ROM, for example) and the name of the vendor or on-line publication service. Because your goal is to help readers retrace your steps if they choose, you can be somewhat flexible in the information you provide. Following are some sample bibliographic entries for electronic publications:

Abstract of a journal article from a database   Longo, N., Langley, S., Griffin, L., Elsas, L. (1995, May). Two mutations in the insulin receptor gene of a patient with leprechaunism [Online]. *Journal of Clinical Endocrinology and Metabolism*, 80 (5). Abstract from: Medline Plus, National Library of Medicine. Item: UI 95263703.

Article from an electronic journal   Olive, K., Scully, S. Big bang nucleosynthesis: An update. [On-line]. *Preprints in Astrophysics*, 27 June 1995. Internet: Electronic Journals. 21 pages. Available: OLIVE@mnhepl.hep.umn.edu.

CD-ROM   *Aquatic Sciences and Fisheries Abstracts* 1988–1994. [CD-ROM] Cambridge Scientific Abstracts. Compact Disc SP-160-010. Available: Silver Platter.

Electronic mail   Awramik, S. (1995, June 28). *Core Course in Science for Humanities Majors* [E-mail to M. Zimmerman], [Online]. Available E-mail mzimmer@humanitas.ucsb.edu.

Computer program   *EndNote Plus and End Link* (IBM PC and Compatibles Version 1.3.2), [Computer Program]. 1994. Available Distributor: Niles & Associates, Inc., Berkeley, CA (Address: 800 Jones St., Zip: 94710).

.

References

Adams, J. 1974. *Conceptual Blockbusting: A Guide to Better Ideas.* San Francisco: W. H. Freeman.

Adewusi, V. A. 1991. Enhanced Recovery of Bitumen by Steam with Chemical Additives. *Energy Sources* 13:121–35.

Aviation Week and Space Technology. 1984. February 13:75.

Bower, D. 1985. Technical Communication in Industrial R&D Groups: A Questionnaire Survey. MIT S.B. Thesis.

Chen, C-C. 1987. *Scientific and Technical Information Sources.* 2nd ed. Cambridge, MA: MIT Press.

Cuadra/Elsevier. 1989. *Online Database Selection: A User's Guide to the Directory of Online Databases.* New York.

Department of Energy. 1991. Safety Evaluation Report: Restart of the K-Reactor, Savannah River Site. Supplement 3. DOE/DP-0093T.

Edelson, R. E., et al. 1979. Voyager Telecommunications: The Broadcast from Jupiter. *Science* 204(4396):913–21.

Engineering Index, Inc. 1992. *Engineering Index Annual.* New York.

Engineering Information, Inc. 1992. *Engineering Index Thesaurus.* Hoboken, NJ.

Fenn, A. J., and G. A. King. 1992. Adaptive Nulling in the Hyperthermia Treatment of Cancer. *Lincoln Lab Journal* 5:223–40.

Gelbart, W. M. 1982. Molecular Theory of Nematic Liquid Crystals. *Journal of Physical Chemistry* 86(22):4298–307.

Gorinson, S. M., et al. The Role of the Managing Utility and Its Suppliers. Staff report to the President's Commission on the Accident at Three Mile Island. Washington, DC: Government Printing Office. Stock No. 052-003-00726-6, Appendix N, p. 227.

Gunning, R., and R. A. Kallan. 1994. *How to Take the Fog Out of Business Writing.* Chicago: The Dartnell Corp.

Guterl, F. 1984. Spectrum/Harris Poll: Education. *IEEE Spectrum* 21(June): 128–34.

Hardt, D. E., et al. 1982. Closed Loop Shape Control of a Roll-Bending Process. *ASME Journal of Dynamic Systems, Measurement, and Control* 104(December): 318–24.

Jagota, A., and P. R. Dawson. 1987. The Influence of Lateral Wall Vibrations and the Ultrasonic Welding of Thin-Walled Parts. *Transactions of the ASME, Series B: Journal of Engineering for Industry* 109(May): 140–46.

Kwack, E. Y., et al. 1992. Morphology of Globules and Cenospheres in Heavy Fuel Oil Burner Experiments. *Transactions of the ASME: Journal of Engineering for Gas Turbines and Power* 114 (April): 338–49.

Merritt, M. W., et al. 1989. Wind-Shear Detection with Pencil-Beam Radars. *The Lincoln Laboratory Journal* 2(3):483–525.

Okerson, A., ed. 1991. *Directory of Electronic Journals, Newsletters, and Academic Discussion Lists.* Washington, DC: Association of Research Libraries.

Pillmoor, J. B., K. Wright, and A. S. Terry. 1993. Natural Products as a Source of Agrochemicals and Leads for Chemical Synthesis. *Pesticide Science* 32:131–40.

Rogovin, M., and G. T. Frampton, Jr. 1980. Three Mile Island: A Report to the Commissioners and to the Public. 2(1):160, 179. Washington, DC: Nuclear Regulatory Commission.

Sugie, E., et al. 1982. A Study of Sheer Crack Propagation in Gas-Pressurized Pipelines. *Transactions of the American Society of Mechanical Engineers* 104: 338–43.

Tufte, E. R. 1983. *The Visual Display of Quantitative Information.* Cheshire, CT: Graphics Press.

United States Presidential Commission on the Space Shuttle Challenger Accident, Report to the President, 1:249–250. Washington, DC: Government Printing Office.

Voight, J. R. 1994. Morphological Variation in Shallow-Water Octopuses (Mollusca: Cephalopoda). *Journal of Zoology* 232:491–504.

Whitney, D. E. 1982. Applying Stochastic Control Theory of Robot Sensing, Teaching, and Long-Term Control. Charles Stark Draper Laboratory. Report No. P-1314.

Index